Postharvest Management and Value Addition of Underutilized Fruit Crops

The Authors

Dr Amit Kumar Singh is currently Teaching Associate, RLBCAU, Jhansi. He completed hisM. Sc. and Ph. D. from BCKVV, West Bengal. He has also been served as Assistant Professor-Horticulture at MS Swaminathan School of Agriculture, CUTM, Gajapati, Odisha for more than two years and received several prestigious awards as well as published dozens of research article in International and National journals.

Dr Arvind Kumar Chaurasiya is currently Assistant Professor, NEHU, Tura Campus, Meghalaya. He did his M.Sc. (Ag.) in Horticulture from NDUAT, Ayodhya and his Ph.D.(Hort) in PHT from BCKVV, West Bengal and received several awards in the field of PHT. He published 2 Articles, 3 National papers, 16 International Papers, One Booklet and 3 Book Chapter.

Dr Awani Kumar Singh is currently Principal Scientist, Vegetable Science in the ICAR-IARI, New Delhi. He has developed 07 high yielding varieties in different vegetables crops and his significant contributions include development of varietyPusa Cherry-1, which is also moderately resistant to root-knot nematode. He has compiled 12 books cum manual chapter, 2 books and published more than 150 articles in national and international journals.

Postharvest Management and Value Addition of Underutilized Fruit Crops

Amit Kumar Singh

Arvind Kumar Chaurasiya

Awani Kumar Singh

2019

Daya Publishing House®

A Division of

Astral International Pvt. Ltd.

New Delhi – 110 002

ISBN: 9789389569162 (Int. Edition)

Publisher's Note:

Every possible effort has been made to ensure that the information contained in this book is accurate at the time of going to press, and the publisher and author cannot accept responsibility for any errors or omissions, however caused. No responsibility for loss or damage occasioned to any person acting, or refraining from action, as a result of the material in this publication can be accepted by the editor, the publisher or the author. The Publisher is not associated with any product or vendor mentioned in the book. The contents of this work are intended to further general scientific research, understanding and discussion only. Readers should consult with a specialist where appropriate.

Every effort has been made to trace the owners of copyright material used in this book, if any. The author and the publisher will be grateful for any omission brought to their notice for acknowledgement in the future editions of the book.

Published by : Daya Publishing House®
A Division of
Astral International Pvt. Ltd.
– ISO 9001:2015 Certified Company –
4736/23, Ansari Road, Darya Ganj
New Delhi-110 002
Ph. 011-43549197, 23278134
E-mail: info@astralint.com
Website: www.astralint.com

रानी लक्ष्मी बाई केन्द्रीय कृषि विश्वविद्यालय
ग्वालियर रोड, झांसी 284 003 (उत्तर प्रदेश) भारत
Rani Lakshmi Bai Central Agricultural University
Gwalior Road, Jhansi 284 003 (U.P.) India

डॉ अरविन्द कुमार
कुलपति
Dr Arvind Kumar
Vice-Chancellor

Phone : 0510-2730777
E-mail : vcrlbcau@gmail.com
Website : www.rlbcau.ac.in

Foreword I

Horticulture has not only become one of the major sectors of the agriculture growth and technological development in India, but also providing employment opportunities across primary, secondary and tertiary sectors, particularly fruit crops, which are relatively climate resilient. Presently, the Horticultural production has overtaken the food grain production in the country and the percentage share of Horticultural crops in value of total agricultural output is nearly thirty percent making India as the second largest producer of fruits in the world. Being first in the production of Banana, Mango, Lime, Lemon and Papaya, a vast production base offers India tremendous opportunities for export and processing due to better connectivity to Europe, the Middle East, Japan, Singapore, Thailand, Malaysia and Korea.

During 2017-18, India exported fresh fruits and vegetables worth Rs. 9,410.81 Crores, out of which fruits comprised of Rs. 4, 229.03 crores. India also exported total processed food worth Rs. 27,257.69 crores in 2017-18. InHorticultural crops, the key processed commodities exported were mango pulp (Rs. 673.92 crores), dried and preserved vegetables (Rs. 944.65 crores), other processed fruits and vegetables (Rs. 3404.70 crores) and alcoholic beverages (Rs. 2105.87 crores). The countries namely UAE, Sri Lanka, Netherland, Bangladesh, Malaysia, Nepal, UK, Saudi Arabia and Qatar were the major importer of fruits and vegetables.

Knowledge management in Horticulture, particularly, fruit crops, shall further help to raise the production through better adoption of modern technologies. The book emphasizes the importance of underutilized fruits in terms of securing nutritional and economic security through utilization of these crops and the value added processed products with the help of photographs and illustrations.

I compliment Dr. Amit Kumar Singh, Dr. Arvind Kumar Chaurasiya and Dr. Awani Kumar Singh for their stupendous effort in bringing out this publication. I am sure that this publication will be of great use by the undergraduate and postgraduate students of Agriculture, Horticulture, Food science, as well as researchers and other stakeholders including industrialists.

(Arvind Kumar)

Camp Office: Room No. 213, KAB-II, Pusa, New Delhi 110 012 *Ph:* (O) 011-25846034

उद्यान एवं वानिकी महाविद्यालय
College of Horticulture and Forestry
रानी लक्ष्मीबाई केन्द्रीय कृषि विश्वविद्यालय, झाँसी, 284003
Rani Lakshmi Bai Central Agricultural University, Jhansi- 284003

डा. ए.के. पाण्डेय, अधिष्ठाता
Dr. A. K. Pandey, Dean

Foreword II

India is the centre of origin for many fruits rich in their medicinal and nutritional properties. In spite of good production, these crops are searching for their commercial success in terms of processing, value addition and export destination. Keeping in view the nutritional and commercial importance of indigenous fruit crops, the well planned and well knitted efforts are required with respect to crop improvement, crop production, post-harvest handling and value addition at Central Institutes, Regional Laboratories and State level institutions. It is gratifying to note that the contributors of the book **"Post Harvest Management and Value Addition of Underutilized Fruit Crops"** have come forward to bring out a comprehensive text on processing and value-addition of lesser known fruits. Apart from describing the nutritional and medicinal importance of underutilized fruits, text gives valuable information on processing and value addition of important underutilized fruits. The book will be very useful to the researchers, students and those who are engaged in processing sectors.

I congratulate authors, Dr. Amit Kumar Singh, Dr. Arvind Kumar Chaurasiya and Dr. Awani Kumar Singh for their effort to bring out this valuable text.

(A. K. Pandey)

Phone No. : 09415983753 E-Mail: pandey.ajail@gmail.com

BIDHAN CHANDRA KRISHI VISWAVIDYALAYA
Department of Post Harvest Technology
FACULTY OF HORTICULTURE

P.O. Krishi Viswavidyalaya
West Bengal, India

Mohanpur-741252, Nadia
Mob: +91 9433513560
E-mail: drsurajitmitra@yahoo.co.in

Dr. Surajit Mitra
Professor & Head

Foreword III

Horticultural produce plays a significant role in human nutrition by supplying vitamins especially vitamin C, minerals, dietary fibre and anti-oxidants to the diet. The quality and safety of underutilized fruit produce reaching the consumer depends upon pre-harvest factors as well as proper post-harvest management practices throughout the chain, from the field to the consumer.

This book reviews the factors which contribute to quality and safety during processing through easy steps of product preparation. Specific examples are given to illustrate the economic implications of investment and applying proper post-harvest technologies. Criteria for the assessment of processing needs, selection of improved technologies to the value addition and for extending appropriate levels of post-harvest information are also discussed. This book will be of great help to faculties, scientists, researchers, students, food processors, academicians, and extension workers and non-governmental organizations (NGOs) who provide training and support to the small-scale processing sectors. It is hoped that the information presented in this book will help in improving post-harvest management practices in developing countries, thereby leading to improvement in the quality and safety of underutilized fruits.

(Surajit Mitra)

Preface

India seems to enjoy a prominent position on the pomological map of the world. The varying weather conditions of this country provide suitable environment for growing a variety of fruits. These fruits are available inabundance and also in different seasons. This has resulted in limited scope for expansion of other minor fruits, though they are nutritious, and are the main source of livelihood for the poor. Fruits, which are among the perishable commodities, are important ingredients in the human dietaries. Due to their high nutritive values, they make significant nutritional contribution to human wellbeing. Most of the underutilized fruits of the tropics are often available only in the local markets and are practically unknown in other parts of the world. A large number of these fruits can grow under adverse conditions and are also known for their therapeutic and nutritive value and can satisfy the demands of the health-conscious consumers. These are the cheaper and better source of the productive foods. If they can be supplied in fresh or preserved form throughout the year for human consumption, the national pictures improve greatly. The perishable fruits are available as seasonal surpluses during certain parts of the year in different regions and are wasted in large quantities due to absence of facilities and know-how for proper handling, distribution, marketing and storage. Furthermore, massive accounts of the perishable fruits produced during particular season result in a glut in the market and become scarce during other seasons. Neither they all be consumed in fresh condition nor sold at economically viable prices.

In the post-green revolution era, even though food grains have been taken care of fruits for want of simple technologies of processing, preservation and transport to various places of need, have suffer postharvest losses, estimated to be more than 25 per cent, and only about 2.2 per cent of the total fruits and vegetables produced are processed. All forms of preserved fruits are in the reach of only the urban elite, and the rural masses that produced more than 90 per cent of the fruits and vegetables are usually derived of their usages. Processing and storage techniques for most underutilized fruit crops are meagerly developed or neglected. Therefore, the quality of the final products cannot be maintained up to required standards. As a result, underutilized fruit trees are not very popular. Improving the processing and storage techniques of underutilized fruits is one of the important aspects in increasing popularity and benefits of underutilized fruit trees. The term underutilized is specified for crops that are not presently cultivated in a particular region or a country, but whose value has been proven elsewhere under similar climatic conditions, and those that are harvested from the wild. However, what is considered underutilized in one area may not be in another area. These fruits may not meet the table values of mango, banana, apple, *etc.*, but they are very popular for their nutritional and medicinal qualities, and can become processed products enhancing their quality and value. Many of these species can tolerate various kinds of abiotic stresses and can be exploited under different situations. However, some of these fruits are not acceptable in the market in fresh form due to their acidic nature and astringent taste. Hence, there is a need to concentrate on research efforts in diversification and popularization of such underutilized fruit crops. To achieve this, there is a need to create demand for such fruit crops in the domestic and international markets. This, to some extent, can be achieved through developing suitable processing and marketing strategies for these underutilized fruits.

On this deliberation, the present book has been attempted which has pursued one vital part in the area on importance, uses and their different value added products of Aonla, Bael, Ber, Carambola, Fig, Jackfruit, Jamun, Karonda, Mahua, and Phalsa. This book brings to attention for educationist thus, seeks to meet the requirements of the Under-graduate and Post-graduate students of fruit science, Horticulture, Postharvest Technology and related areas on the one hand and to those, who desire to take up value addition of underutilized fruit crops in other hand.

In this book, published information has been freely used. Some information has been collected from electronic media/website, the sources of which duly been acknowledge. I would like to record my special thank to the Chancellor and Vice-Chancellor of the Rani Lakshmi Bai Central Agricultural University, Jhansi to encourage and facilitate for making this valuable book. Lastly, the author looks forward to receive valued opinion and suggestions from the readers so as to bring out the book in a better form in future.

Authors

Contents

Introduction

The green revolution and subsequent efforts through the application of science and technology for increasing food production in India have brought self-reliance in food. The impetus given by the government, Central and State Agricultural Universities, State Departments of Agriculture and other organizations through the evolution and introduction of numerous hybrid varieties of cereals, legumes, fruits and vegetables and improved management practices have resulted in increased food production. However, the nation still faces the problem of the use of improper methods for storage of foodstuffs, leading to great wastage of the food produced. Such losses in the food front aggravate the existing syndrome of under nutrition and malnutrition. Fruit and vegetables, which are among the perishable commodities, are important ingredients in the human dietaries. Due to their high nutritive values, they make significant nutritional contribution to human wellbeing. They are the cheaper and better source of the productive foods. If they can be supplied in fresh or preserved form throughout the year for human consumption, the national pictures improve greatly. These perishable fruits are available as seasonal surpluses during certain parts of the year in different regions and are wasted in large quantities due to absence of facilities and know-how for proper handling, distribution, marketing and storage. Furthermore, massive accounts of the perishable fruits produced during particular season result in a glut in the market and become scarce during other seasons. Neither they all be consumed in fresh condition nor sold at economically viable prices.

In the post-green revolution era, even though food grains have been taken care of fruits for want of simple technologies of processing, preservation and transport to various places of need, have suffer Postharvest losses, estimated to be more than 25 per cent and only about 2.2 per cent of the total fruits and vegetables produced are processed. All forms of preserved fruits are in the reach of only the urban elite, and the rural masses that produced more than 90 per cent of the fruits and vegetables are usually derived of their usages. Even in developing countries, agriculture is the mainstay of the economy. As such, it should be no surplus that agriculture industries and related activities can account for a considerable proportion of their output. Of the various types of activities that can be termed as agriculturally based fruit processing are among the most important. Therefore, fruit processing has been engaging the attention of planners and policy makers as it can contribute to the economic development of rural population. The utilization of resources both material and human are one of the ways of improving the economic status of family. India is the second largest producer of food next to China and is booming with a growth rate of 20 per cent a year. It has the potential of being the biggest with the food and agricultural sector contributing around 26 percent of the Gross Domestic Product (GDP). Key to giving a fillip to India's food production would be minimization of all the wastages in the food distribution and tapping the myriad possibilities that biotechnology and functional genomics offer. Wild varieties of plants, yielding edible fruits growing throughout the Himalayas, contributed directly to cultural heritage of India. Even today, these fruits are eaten in plenty by local people, as they are commonly available in abundance in their habitats.

What is "Underutilized Fruits?"

The term underutilized is specified for crops that are not presently cultivated in a particular region or a country, but whose value has been proven elsewhere under similar climatic conditions, and those that are harvested from the wild. However, what is considered underutilized in one area may not be in another area. These fruits may not meet the table values of mango, banana, apple, *etc.*, but they are very popular for their nutritional and medicinal qualities, and can become processed products enhancing their quality and value includes in human diet. Many of these species can tolerate various kinds of abiotic stresses and can be exploited under different situations. "Fruit crops, which are neither grown commercially on large scale nor traded widely, are called underutilized fruits crops. These are cultivated, traded and consumed locally." Underutilized fruit crops have many advantages in terms of easiness to grow, hardy in nature and production of good crop even under adverse conditions. Underutilized fruit crops play crucial role in successful running of processing industry round the year in the region. A large proportion of rural population depends on locally available fruits to meet their datary requirements. These fruit crops have their own history of consumption, local people are well aware their nutritional and meditational properties.

What is "Value Addition?"

"Value addition" means adding value to a raw product by taking it to, at least, the next stage of production. This can be as simple as retaining ownership of your calves and wintering them on wheat pasture or placing them in a feedlot. Adding value may be as elaborate as going all the way to the consumer with a "case-ready" food product.

India has broad spectrum of agro-climatic regions of fruits trees and most of the area still lies under indigenous fruit crops, which are neglected. These fruits are major source of the dietary fiber, minerals, vitamins, carbohydrates, proteins, and fats. These underutilized fruits are also rich in calcium, phosphorus, iron and magnesium. They are also a good source of the vitamin C, vitamin A and thiamine, riboflavin, niacin, pantothenic acid and folic acid. These fruits contain good amount of dietary fiber, essential for bowel movement and possibly for the prevention of several diseases such as appendicitis, colon cancer, diabetes, gallstones, obesity *etc.* Postharvest value addition includes primary, secondary, and tertiary processing, operations performed on farm produce. It is to provide longer shelf life, maintain/ improve quality, and enhance form, space and time utility of the produce for food, feed, fiber, fuel and industrial purposes. The Postharvest operations include on-farm handling, cleaning, grading, moisture conditioning, milling, extraction, cooling, freezing, roasting, puffing, flaking, retort processing, packaging, transport and storage.

This book consists ten underutilized fruit crops (Aonla, Bael, Ber, Carambola, Fig, Jamun, Jackfruit, Karonda, Phalsa and Mahua) their importance, uses, health benefits, harvesting, yield, maturity stage, grading, storage and their value-added products with their recipe and uses. Processing and storage techniques for most underutilized fruit crops are weakly developed or neglected. Therefore, the quality of the final products cannot be maintained up to required standards. As a result, underutilized fruit trees are not very popular. Improving the processing and storage techniques of underutilized fruits is one of the important aspects in increasing popularity and benefits of underutilized fruit trees.

1

Aonla

Scientific Name: *Emblica officinalis* Gaertn

Family: Euphorbiaceae

Edible portion: Mesocarp and Exocarp

Aonla (*Emblica officinalis* Gaertn., family- Euphorbiaceae) is an important fruit crop grown in tropical and subtropical parts of India, China, Indonesia and the Malay Peninsula. It is rich source of vitamins, tannins with high nutritional and medicinal values. Generally Aonla is considered as "Wonder Fruit for Health". It is known by different names like amla, amalakki, nelli and Indian gooseberry. It had been found in dry deciduous forests of India. It is a highly valued among indigenous medicines. The edible tissue of Aonla contains about three times more protein and 160 times more vitamin C as compare to apple (Barthakur and Arnold, 1991). The high acceptability of Aonla fruits could be due to their high medicinal values and the fact that they are richest source of vitamin C among all other fruits except for Barbados cherry (Asenjo, 1953; and Shankar, 1969). Being a very rich source of vitamin C (500-1500 mg/100g) (Shankar, 1969) and other nutrients like polyphenols, pectin, iron, calcium and phosphorus (Nath *et al.,* 1992) the fruit has a potent antioxidant property (Frei *et al.,* 1989). It has an astringent taste and is, therefore, not popular as a table fruit. However, it shows great potential for processing into various quality products, which can have great demand in national as well as international market.

Importance and Use

All parts of the plant are used for medicinal purpose. The fresh or dry fruit is used in traditional medicines for the treatment of diarrhea, jaundice and inflammations (Deokar, 1998). The pulp of the fruit is smeared on the head to dispel headache and dizziness (Perry, 1980). Their leaves and fruit have been used for fever and inflammatory treatments by rural populations in its growing areas. The earlier study have demonstrated potent anti-microbial (Ahmad *et al.*, 1998), anti-oxidant, adaptogenic (Rege *et al.*, 1999), hepatoprotective (Jeena *et al.*, 1999), anti-tumour (Jose *et al.*, 2001) and anti-ulcerogenic activities (Sairam *et al.*, 2002) in the fruits. Leaf extracts have been shown to passes anti-inflammatory activity (Asmawi *et al.*, 1993; Ihantola-Vormisto *et al.*, 1997). Ethanol and aqueous extract of *Emblica officinalis* has shown significant anti-inflammatory activity (Sharma *et al.*, 2003). Aonla fruits are widely used in the Unani and Ayurvedic systems of the medicines in the forms of the fresh fruit or powder in various preparations such as Chyawanprash, Rasayana, Triphala and Aristha which promotes health and longevity (Rajkumar *et al.*, 2001). The fruit is acrid, cooling, refrigerant, diuretic and laxative, hence used for treating chronic dysentery, bronchitis, diabetes, fever, diarrhea, jaundice, dyspepsia, and coughs *etc.* The gallic acid present in Aonla fruit has antioxidant properties. In addition to this potent antioxidant several active tannoid principles (Emlicannin A, Emblicannin B, Puniglucon and Pedunculagin) have been identified which appear to account for its health benefits (Rastogi, 1993; Rao *et al.*, 1985). Amla has been reported to posses expectorant, purgative, spasmolytic, antibacterial, hypoglycemic (Jamwal *et al.*, 1959; Jayshri and Jolly, 1993), heaptoprotective and hypolipidemic (Thakur and Mondal, 1984) activity. The aqueous extract has been reported to have anti-pyretic laxative and tonic properties and also showed antibacterial activity (Vinayagamoothy, 1982). Amla was used to prepare ready-to-serve beverage (Deka *et al.*, 2001), candy, powder, pickle (Tripathi *et al.*, 1988; Nath and Sharma, 1988; Chauhan *et al.*, 2005), preserve, juice, shreds, dried powder (Tandon *et al.*, 2003a) *etc.* Owing to its nutritive and miraculous medicinal properties, this fruit has acquired wide popularity. The fresh fruits are generally not consumed due to their high astringency but it has got great potential in processed forms (Ranote and Singh, 2006).

Composition

Food Value/100g of Edible Portion

Characters	*Composition*
Moisture	81.8g
Carbohydrate	13.7g
Protein	0.5g
Fat	0.1g
Fiber	3.4g
Minerals	0.5g
Calcium	50 mg
Phosphorus	20 mg
Iron	1.2 mg
Carotene	9 mg

Characters	Composition
Thiamine	0.03 mg
Riboflavin	0.01 mg
Niacin	0.2 mg
Vitamin C	600 mg
Choline	256 mg

Source: Ganachari ***et al.,*** 2005

Harvesting and Chemical Treatments

The change in seed colour from creamy white to brown is indicative of fruit maturity. Fully developed fruits, which show sign of maturity, are harvested. Delay in harvesting results in heavy dropping of fruits particularly in varieties like Banarasi and Francis. It also adversely affects the following years bearing. Individual fruits are plucked by climbing on tree with the help of pegged bamboo or ladder. Harvesting should be done in early or in the late hours. There is a linear increase of quality parameters from 35 days old to fully matured fruit (120 days) of Aonla and hence it is ideal to harvest fruits at 120 days after fruit set particularly for processing (Sivakumar and Sundaram, 2010).

Chemical treatments play an important role in increasing the shelf life of Aonla. Pre-harvest sprays (twice) of 1 per cent calcium nitrate + 0.1 per cent Topsin-M decreased the weight loss (11.09 per cent) and decay loss (14.43 per cent) to the extent of 11.09 per cent and 14.43 per cent and prolonged shelf-life up to 20 days compared with 10 days in the control at ambient temperature (Yadav and Singh, 1999). Further treatment with Topsin-M and Bayleton controlled *Penicillium oxalicum* for 10 days and *Aspergillus niger* for 20 days and extended shelf life (Yadav and Singh, 1999). Singh *et al.* (2010) found that fruits treated with 1.5 per cent $CaCl_2$ and stored in ZECC (zero energy cool chamber) recorded least PLW (16 per cent), spoilage loss (16.5 per cent), respiratory activity (83 mg CO_2/kg/hrs.) and exhibited 11 days of shelf-life, while untreated (control) had 6 days economic life. It was closely followed by 1 per cent $CaCl_2$ + ZECC treatment.

Postharvest treatment with Calcium nitrate (1 per cent) minimized the weight loss during the storage period and no pathological loss was observed with Borax (4 per cent) up to 9 days of storage (Nath *et al.,* 1992). Postharvest treatment with 6 per cent Waxol + 1 per cent $Ca(NO_3)_2$ followed by 6 per cent Waxol + 400 ppm CCC recorded lower PLW and moisture loss throughout the storage period. The treatment with 6 per cent Waxol + 0.1 per cent carbendazim was found effective in reducing decay loss. Treatment with 6 per cent Waxol + 100 ppm GA_3 resulted in maximum retention of ascorbic acid followed by treatment 6 per cent waxol + 400 ppm CCC (Dhumal *et al.,* 2008). Patel and Sachan (1995) used calcium nitrate (1 per cent), GA_3 (40 ppm), CCC (400 ppm) and kinetin (10 ppm) for experiment. After treatment they packed fruits in perforated polythene bags and stored at ambient temperature. The PLW and rotting percent increased with the increase in storage period. According to them calcium nitrate (1 per cent) was the best treatment to minimize the weight loss of fruits. No rotting was observed up to 9 days of storage in kinetin (10 ppm) treated fruits. GA_3 (40 ppm) treatment gave better retention of vitamin 'C' during storage of Aonla fruits.

Packaging

At present proper packaging is inadequate in case of Aonla, which can be packed in gunny bags of 50 to 100 kg capacity. But problem is that the fruits got impact, vibration and compression injuries during transportation in these gunny bags. The corrugated fiber boxes are better as this provides appropriate atmosphere and ventilation inside the box, printable information at low cost and recyclable also. Newspaper lining should be provided inside the corrugated fiber board (CFB) cartons. Minimum spoilage (16.0 per cent) was noticed in corrugated fiber board boxes with newspaper liner package followed by CFB boxes with polythene liner (17.0 per cent), where as it was highest in gunny bag without any liner (30.19 per cent) after 13 days of storage (Singh *et al.,* 2005). Singh *et al.* (1993) found that wooden crate with polythene liner is most suitable for packing and long distance transportation of Aonla fruits. Percent weight loss and bruising were minimum in this container as compared to gunny bag. According to Dhumal *et al.* (2008) packaging of fruits in perforated PE bags had remarkable effect in reducing PLW, retention in moisture and acceptable physico-chemical qualities. The shelf life of Aonla fruits was extended up to 15 days at room temperature when fruits were treated with 6 per cent Waxol + 400 ppm CCC and packed in perforated PE bags. This treatment combination recorded the maximum score for marketability. The physico-chemical changes were faster in untreated fruits (control) packed in nylon net bags.

Storage

Aonla fruits are highly perishable in nature as it's storage life in atmospheric conditions after harvesting is very limited. Storage facilities such as cold storage and controlled/modified atmospheric storage are very expensive and not in the direct reach of poor farmers (Kumar and Nath, 1993). A big problem in Aonla fruit storage has been observed by Premi *et al.* (1999), which are known as white speck (WS). They treated Aonla fruits with using salt, KMS, $CaCl_2$, sodium benzoate and acetic acid with a single and a combination of different concentration and they found that white speck problem which found during storage get maximum retention with a chemical solution containing 10 per cent salt and 0.04 per cent KMS. Aonla fruits treated with 1 percent solution of $CaCl_2$ and 1 percent solution of $CaNO_3$ for 30 min. each, and with single layer of malted paraffin wax, congealing point of 58-60 ^0C was found that the physically injured fruits with wax coating for three weeks without adversely affected the length and width ratio, physiological loss in weight, total soluble solids and vitamin C (Pathak *et al.,* 2009). Singh and Kumar (2000) studied the postharvest treatment effect on Chakaiya cultivar of Aonla and they dipped fruits up to 10 min. in an aqueous solutions of gibbrellic acid (10 and 25 ppm) and kinetin (100 and 150 ppm) then fruits were surface dried before packing and kept it up to 24 days with minimum loss in chlorophyll and carotenoids. Singh and Kumar (1997) reported that the decay loss was minimum (26.56 per cent) in modified storage condition on 24th day of storage whereas it was maximum (48.70 per cent in zero energy cool chamber. The fruits may be kept in cold storage for 7-8 days at 0-2 ^{0}C and 85-90 per cent relative humidity.

Screening of Aonla for Processing

Singh *et al.* (2004) evaluate Aonla varieties for fruit processing and studied on five Aonla cultivars *viz.* NA-6, NA-7, NA-10, Chakaiya and Kanchan on the basis of physic-chemical composition and found that the Aonla variety NA-6 showed better attributes for processing of various products of economic importance. Nath and Sharma (1998) also screened out the cultivar for processing. They were taken five Aonla cultivars *viz.* Banarasi, Chakaiya, Francis, Krishna and Kanchan and found that cultivars differ for their processing quality. Cultivar Chakaiya found suitable for beverages (nectar, squash and syrup) and jams were as Banarasi proved its suitability for candy and pickle preparation.

Value Added Products

Although Aonla fruit is highly nutritive with great medicinal values, research in the direction of value addition for better utilization of the product is limited. Attempts are being made to produce products, which are not only nutritionally delicious but also acceptable among the consumers such as murabba (preserve), pickle, candy, juice, squash, jam, jelly, powder, *etc.* are prepared from Aonla fruits (Tripathi *et al.*, 1988; Nath and Sharma, 1998). Low cost preservation technology was applied to prepare products by dehydration as dried whole fruit, flakes (Verma and Gupta, 2004), slice (Alam *et al.*, 2010), supari (Damame *et al.*, 2002), shreds (Sagar and Kumar, 2006), and powder (Sharma *et al.*, 2002; Alam and Singh, 2005; Vijayanand *et al.*, 2007). Blanching with hot water or with potassium metabisulphite (KMS) before drying checks the enzymatic spoilage and also improves the colour and texture of the shreds (Prajapati *et al.*, 2009).

The prospect of utilization of blended RTS of Aonla and bael fruits at the ratio of 25: 75 (Ram *et al.*, 2011) and blended Aonla juice with grape juice at the ratio of 20:80 on the basis of overall sensory quality and vitamin C content (Jain and Khurdiya, 2004) has been reported. Small size Aonla fruits, which are not suitable for preparation of preserve and other confectionary items, may be utilized for pickle making (Goyal *et al.*, 2008).

Aonla Jam

Jam is a concentrated fruit product processing a fairly heavy body rich in natural fruit flavor. Pectin in fruit gives it a good set and high concentration of

Technological Flow-Chart for Processing of Jam

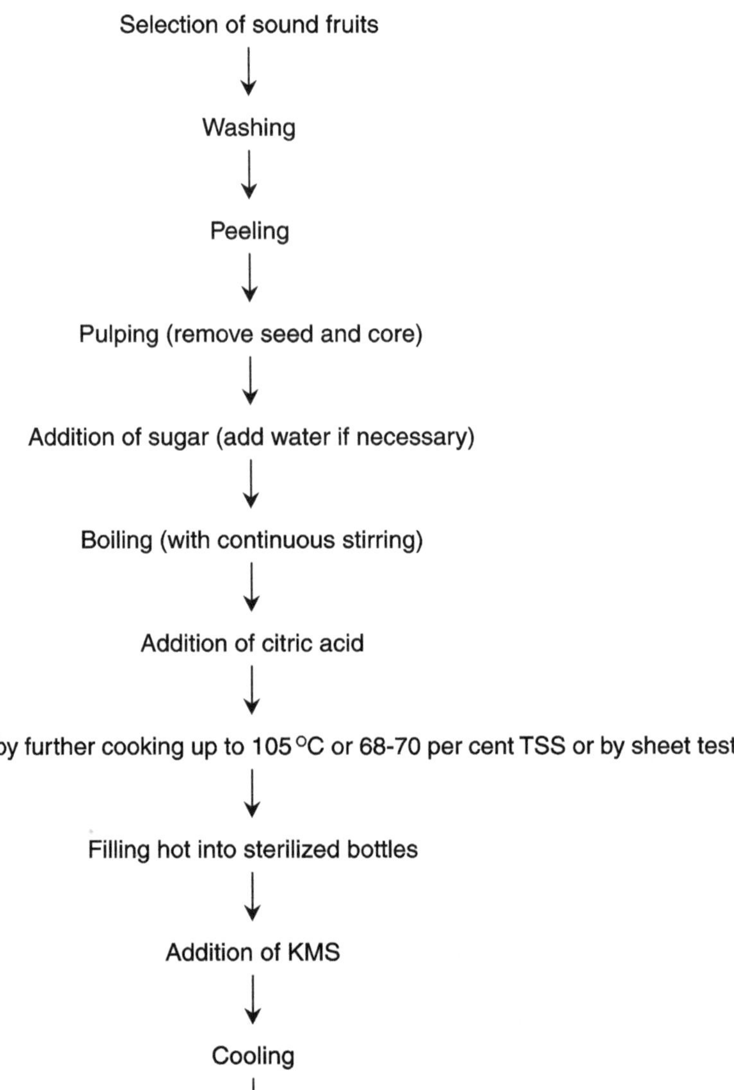

sugar facilitates its preservation. It is prepared by boiling the fruit pulp and juice with sufficient quantity of sugar to a reasonably thick consistency to hold the fruit tissues in position. A fruit jam should contain 45 per cent of fruit portion and 68 per cent of total soluble solids (Pareek *et al.*, 2011).

Sl.No.	*Ingredients*	*Quantity*	*Sl.No.*	*Ingredients*	*Quantity*
1	Aonla (pulp)	1 kg	4	KMS	40 ppm
2	Water	150 ml	5	Pectin	1 g
3	Sugar	750 g			

Aonla Sauce

Sauce and ketchup have some ingredients and are also cooked in the same manner as chutney except that fruit/vegetable pulp or juice used is sieved after cooking to remove skin, seed and stalk of fruits, vegetables and spices to give a smooth consistency to the final product. Cooking a sauce/ketchup takes longer time than chutney because in case of sauce/ketchup, fine pulp or juice is used. Five kg of sauce containing 50 per cent Aonla pulp and 50 per cent tomato pulp with 75 g sugar, 10 g salt, 60 g onion, 6 g garlic, 12 g ginger, 5 g red chilies and 12 g hot spices was prepared. Acetic acid and sodium benzoate as preservative were added @ 1ml and 0.3g/kg of final product, respectively. Finally the sauce was filled in glass bottles and crown corked fallowed by processing in boiling water for 30 minutes, and air-cooled. The product was highly acceptable even after the storage period of more than 9 months (Ranote and Singh, 2006).

Sl.No.	*Ingredients*	*Quantity*	*Sl.No.*	*Ingredients*	*Quantity*
1	Aonla (pulp)	1 kg	7	Onion	50 g
2	Tomato (pulp)	1 kg	8	Garlic	5 g
3	Water	200 ml	9	Ginger	10 g
4	Sugar	75 g	10	Chilli Powder	5 g
5	Salt	10 g	11	Cinnamon, cardamom, aniseed, cumin	10 g each
6	Vinegar	25 ml	12	Sodium Benzoate	0.25g/kg of finished products

Technological Flow-Chart for Processing of Sauce (Aonla)

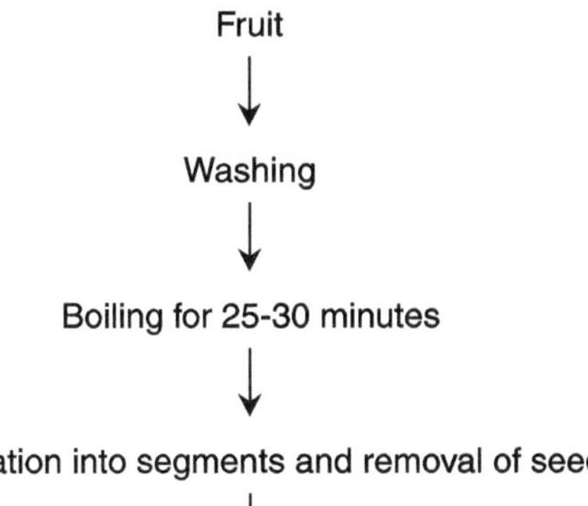

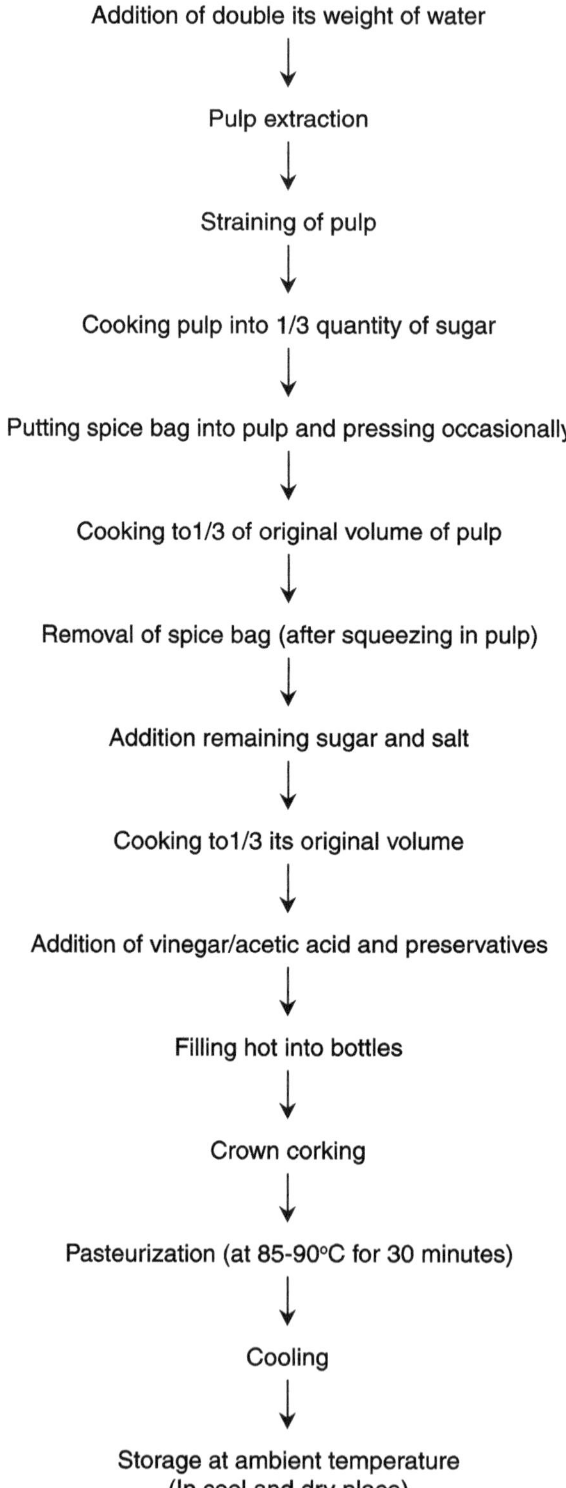
Addition of double its weight of water
Pulp extraction
Straining of pulp
Cooking pulp into 1/3 quantity of sugar
Putting spice bag into pulp and pressing occasionally
Cooking to1/3 of original volume of pulp
Removal of spice bag (after squeezing in pulp)
Addition remaining sugar and salt
Cooking to1/3 its original volume
Addition of vinegar/acetic acid and preservatives
Filling hot into bottles
Crown corking
Pasteurization (at 85-90°C for 30 minutes)
Cooling
Storage at ambient temperature
(In cool and dry place)

Aonla Chutney

Aonla chutney is generally hot, sweet, smooth spicy, mellow flavoured and appetizing. Sometimes raising and dry fruits are also added to increase its taste and nutritional value.

Sl.No.	*Ingredients*	*Quantity*	*Sl.No.*	*Ingredients*	*Quantity*
1	Aonla (pulp)	1 kg	3	Sugar	1.25 kg
2	Water	750 ml	4	Citric acid	2-3 g

Technological Flow-Chart for Processing of Chutney (Aonla)

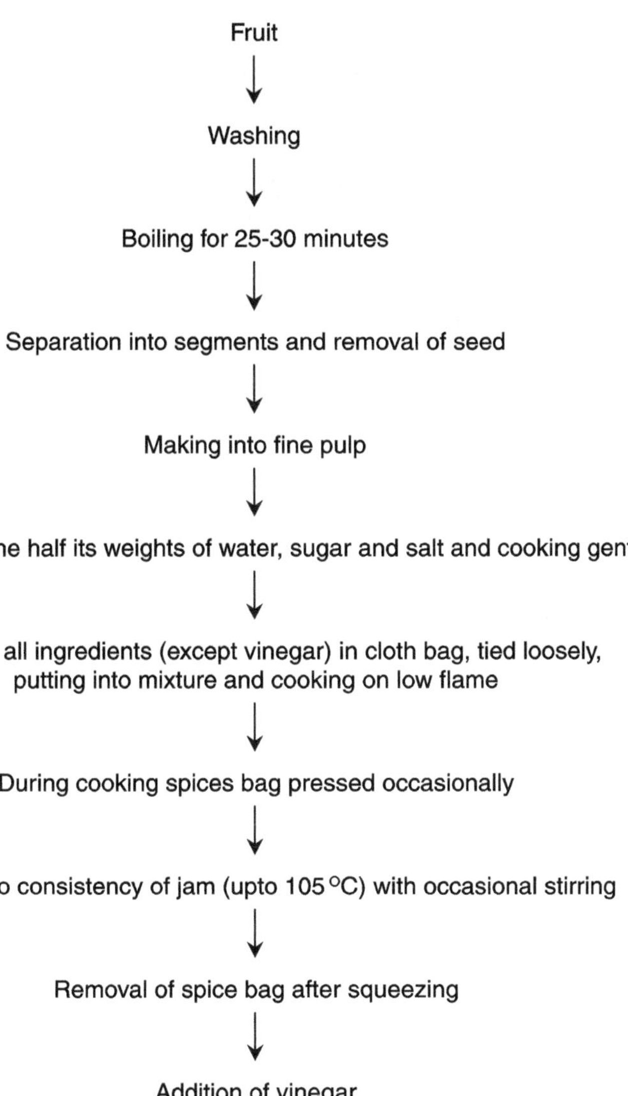

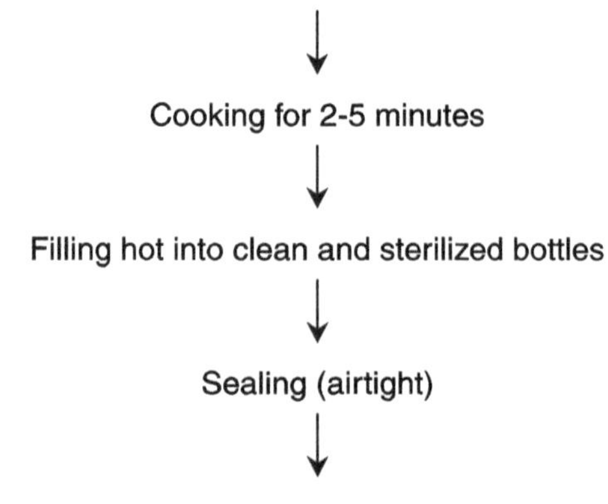

Aonla Preserve

Aonla preserve is an extremely popular traditional product, which is also known as Amla Murabba in India. It has the beneficial effect of purifying blood and also helps in reducing the cholesterol level and improving eyesight (Ranote and Singh, 2006). Preserve are prepared from matured, whole or in large pieces of fruit, in which sugar are impregnated till it becomes tender and transparent. Minimum fruit portion in preserve should be 55 per cent. Pricking (piercing) is done in Aonla to remove the astringency and to allow the syrup to go inside the fruits. Sahu *et al.* (2010) studied on Aonla preserve which treated from alum found best as compare to water, salt, lime and ethephon, and Total Soluble Solids, Total Sugar, Reducing Sugar, Acidity were found increasing in order while, ascorbic acid and organoleptic value were found decreasing in order during storage of Aonla preserve. Geetha *et al.* (2006) Aonla fruits treated with salt solution in 2-4 per cent for three days and then blanched in boiling alum solution for 4 min. during preserve preparation.

Sl.No.	*Ingredients*	*Quantity*	*Sl.No.*	*Ingredients*	*Quantity*
1	Aonla (pulp)	1 kg	4	Citric acid	1.5 g
2	Alum/Salt	20 g	5	KMS	1.2 g
3	Water	1 ltr			

General Considerations

1. Prick with fork, needle or goose berry pricker.
2. Steep in 2 per cent salt solution for 24 hrs to remove astringency
3. Wash and dip in 2 per cent alum solution for 24 hours then wash thoroughly
4. Blanch until soft but segments do not break or crack.
5. Cooking of amla directly in syrup causes shrinking of fruit and reduces absorption of sugar.

6. Therefore, the fruit should be blanched first to make it soft enough to absorb water, before steeping in syrup.

Fruits may be cooked in syrup by two processes as given below:

1) Rapid Process

1. Amla are cooked in a low sugar syrup. Boiling is continued with gentle heating until the syrup becomes sufficiently thick.
2. Rapid boiling should, however be avoided as it makes the fruit tough, especially when heating is done in a large shallow pan with only a small quantity of syrup.
3. The final concentration of sugar should not be less than 68 per cent which corresponds to a boiling point of 106 ^{0}C.
4. This is a simple and cheap process but the flavour and colour of the product are lost considerably during boiling.

2) Slow Process

1. The amla is blanched until it becomes soft.
2. Sugar, equal to the weight of fruit, is then added to the fruit in alternate layers and the mixture allowed to stand for 24 hrs.
3. During this period, the amla gives out water and the sugar goes into solution, resulting in a syrup containing 37-38 per cent TSS.
4. Next day, the syrup is boiled after removal of amla to raise its strength to about 60 per cent TSS.
5. A small quantity of citric acid (1 to 1.5 g/kg sugar) is also added to invert a portion of the cane sugar and thus prevent crystallization.
6. The whole mass is then boiled for 4-5 min. and kept for 24 hrs.
7. On the third day, the strength of syrup is raised to about 65 per cent TSS by boiling.
8. The fruit is then left in the syrup for a day.
9. Finally, the strength of the syrup is raised to 70 per cent TSS and the amla are left in it for a week.
10. The preserve is now ready and is packed in containers. This method is usually practiced.

Aonla Candy

Fruit candies are becoming more and more popular because of high acceptability, minimum volume, higher nutritionally value and longer storage life. These have additional advantage of being least thrust provoking and ready to eat snacks (Ranote and Singh, 2006). A fruit impregnated with sugar, removed drained and dried is called as candied fruit or fruit candy. It should have maximum

75 per cent of total soluble solids. For the preparation of Aonla candy, mature fruits are washed, pricked and dipped in 2 per cent salt solution for 24 hours. Then fruits are washed and dipped in 2 per cent alum solution for 24 hours. The fruit are thoroughly washed and blanched in boiling water for 5 minutes and steeped in 50 ^{0}Brix syrup solutions for 24 hours. The next day steeping is done in 60 ^{0}Brix for 24 hours. Again steeping is done in 70 ^{0}Brix for 72 hours. Excess syrup is drained. The fruit are dried to 15 per cent moisture content and coated with powdered sugar/pectin. Packaging is done in polythene pouches (400 gauge) by Ranote and Singh, 2006.

Sl.No.	*Ingredients*	*Quantity*	*Sl.No.*	*Ingredients*	*Quantity*
1	Aonla (pulp)	1 kg	4	KMS	1.2 g
2	Sugar	1.12 kg	5	Citric acid	6.5 g
3	Water	500 ml			

Technological Flow-Chart for Processing of Preserve and Candy

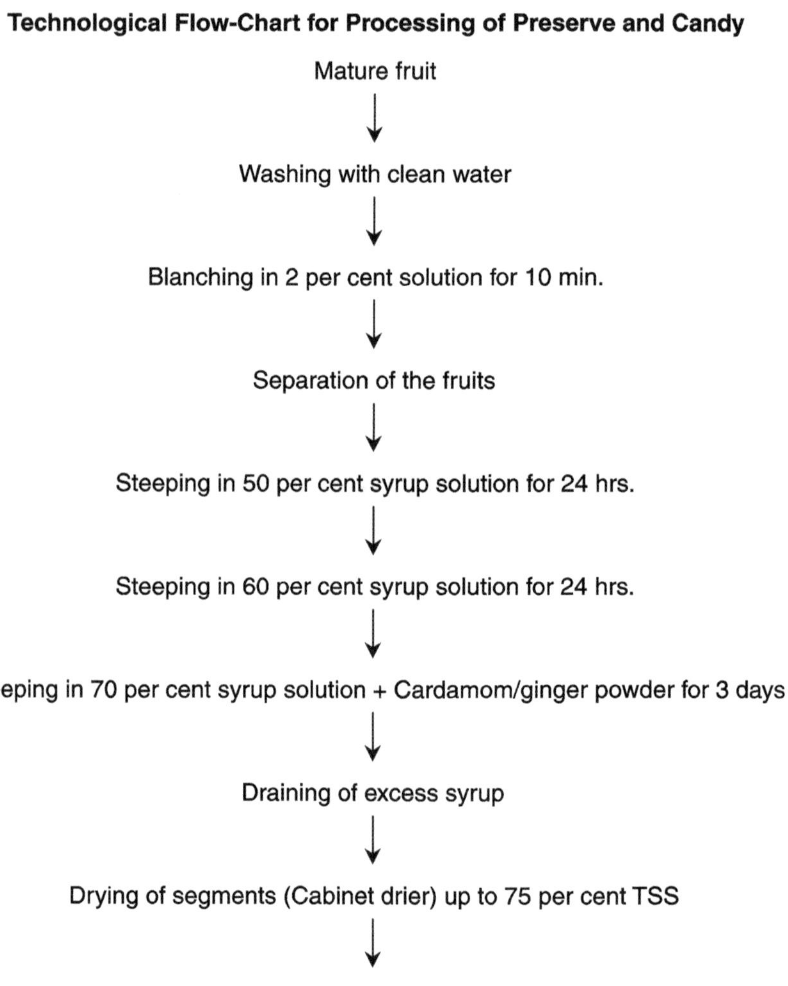

Wrapping with powdered sugar and mixing pieces with fried spices

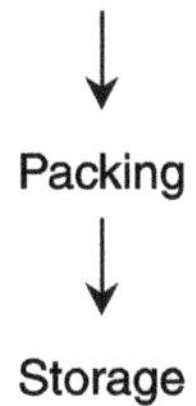

Packing

↓

Storage

(Nayak *et al.*, 2012)

Blended Ready-to-Serve Beverages

Ram *et al.* (2011) used Aonla and bael fruits for blended RTS (ready-to-serve) beverages. The blended RTS beverage prepared with 25 per cent Aonla and 75 per cent bael (*Aegle marmelos*) pulp with 15 ^{0}Brix TSS and 0.25 per cent titratable acidity was similar to the beverage prepared with only bael pulp; it was better than pure Aonla beverages and found that increased concentration of Aonla pulp decreased the acceptability of the beverages. Jain and Khurdiya (2004) conducted an experiment on Aonla juice blended with apple, lime, pomegranate, Perlette grape and Pusa Navrang grape. When gooseberry juice was blended with other fruit juices for the preparation of ready-to-serve (RTS) beverages, it boosted their nutritional quality in terms of vitamin C content. The authors found that the blending of Aonla juice and grape juice [20:80] gave the best result on the basis of overall sensory quality and vitamin C content. Chandan *et al.* (2010) prepared an Aonla RTS beverage with drained Aonla syrup. They obtained the drained syrup from blanched slices of Aonla steeped in salt for 2 hrs., followed by steeping in 70 ^{0}Brix syrup for 24 hrs., and adjusted it to 20 ^{0}Brix containing 2 per cent lime juice + 1 per cent ginger juice, which was found to be acceptable, with good organoleptic scores.

Technological Flow-Chart of RTS

Extraction of juice

↓

Preparation of syrup solution (sugar + water + acid)

↓

Cooling syrup upto room temperature

↓

Mixing juice and syrup

↓

Homogenization

↓

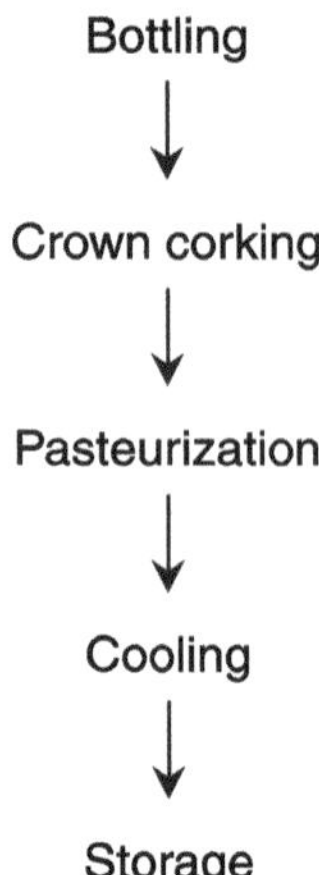

Osmo-Air Dried Aonla Slices

This is a new product in which Aonla fruits are dipped in sugar syrup of 70 per cent concentration over night and the hot air dried in an oven at 60 °C for a day.

Sl.No.	Ingredients	Quantity	Sl.No.	Ingredients	Quantity
1	Aonla (shred)	1 kg	3	Water	750 ml
2	Sugar	1.5 kg	4	Citric acid	2 g

Osmo-Vacuum Dehydrated Aonla Segment

Quality improvement is not related only to the water removal without thermal stress but also to the impregnated solutes. With the correct choice of solutes and a controlled and equilibrated ratio of water removal and impregnation, it is possible to retain natural flavor and colour in fruit products (Barat at al. 1998; Pokharkar *et al.,* 1997). To overcome the astringency problem in Aonla, blanched the fruits in boiled alkali (2 per cent NaOH) solution for 5-8 min. Lye treated fruits were washed thrice in tap water and then soaked in citric acid (0.1 per cent) for 20 min. to neutralize alkali treatment and to remove astringency (Kumar and Sagar, 2012; Suresh and Sagar 2009). The blanched fruits were dipped in cold water for 2 min. for easy separation of segments manually and removal of seeds, weighed amount of Aonla segments were suspended in 60 per cent sugar solution containing 0.05 percent potassium meta-bisulphite and 0.1 per cent citric acid in a stainless vessel. The temperature (60 °C) and sugar concentration (60 °Brix) of the solution were maintained at pre set value. The ratio of fruits and osmotic solution was maintained at 1:4 to ensure proper soaking of samples. Samples were withdrawn from osmotic solution after 6 hours of immersion, drained quickly and wiped gently with tissue paper to remove the adhering sugar solution from surface of the segments. The pretreated samples were spread on perforated aluminum tray load of 0.40 g/ cm^2 and were kept in vacuum drier 640 mmHg up to 9 per cent moisture of final product, then it air packed and stored at ambient temperature and 55-65 per cent relative humidity (Suresh and Sagar, 2009).

Aonla Juice

Jain and Khurdia (2009) studied on ascorbic acid content and non-enzymatic browning in stored Indian gooseberry juice as affected by sulphitation and storage. Juice was extracted by crushing and pressing the blanched fruits (after seed removal) and water 1:1 ratio and treated as A) pasteurization at 90 ^{0}C for 1 min. and filling hot in pre-sterilized, hot glass bottles (Jain *et al.,* 2003), B) treatment 300 ppm SO_2 (KMS) and C) pasteurization at 90 ^{0}C for 1 min., cooled to 60 ^{0}C and 350 ppm SO_2 was added before sealing in glass bottles and they found after 6 months of storage, maximum vitamin C content (Cv. Chakaiya) 232.7 mg/100 ml. was observed in SO_2 treated juice stored at low temperature.

Sl.No.	*Ingredients*	*Quantity*
1.	Aonla (pulp)	1 kg
2.	Water	1 ltr
3.	KMS	6.5 g

Technological Flow-Chart for Processing of Aonla Juice

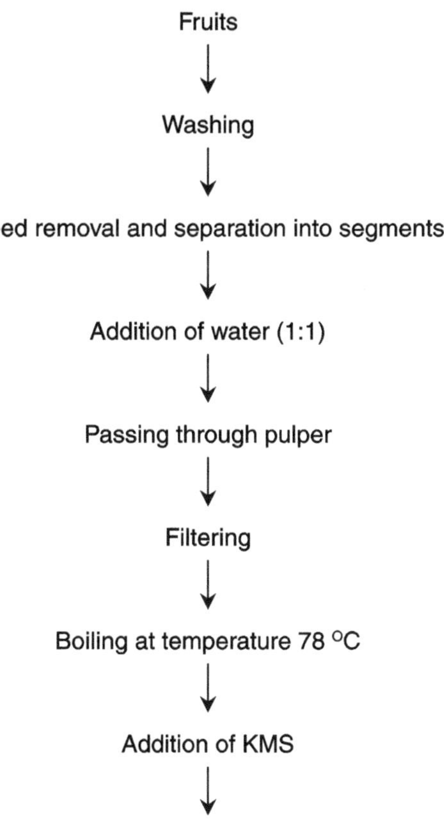

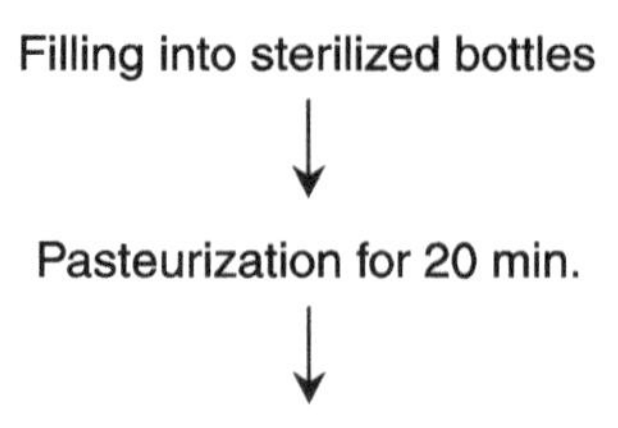

Aonla Powder

Mishra *et al.,* 2009 has been studied on Aonla powder from cultivar Chakaiya. Aonla fruits were cleaned and cut into pieces, and immediately pressed to obtain juice using a small laboratory manual press. Juice were dried in five types of drier *viz.* freeze drying, sun drying, vacuum drying, spray drying and tunnel drying fallowed by grinding in a mixer grinder and filtration using muslin cloth. In all five methods they found that the freeze drying (Model Alpha 1-4, Martin Christ, Germany at 16 ^{0}C for 16 hours) was best in respect of high retention of calcium, phosphorus, iron, ascorbic acid and colour fallowed by spray drier.

Aonla Peels

Aonla peels are prepared by drying Aonla pulp mixed with ground cumin, ginger and sends salt. These peels are very delicious, digestive and rich source of vitamin C.

Sl.No.	*Ingredients*	*Quantity*	*Sl.No.*	*Ingredients*	*Quantity*
1	Aonla (pulp)	1 kg	3	Ginger chopped	50 g
2	Salt	30 g	4	Cumin powder	30 g

Aonla Pickle

The preservation of fruit or vegetables in common salt or vinegar is called pickling. Spices and oil may also be added in pickle. Common salt @ 15 per cent prevents its spoilage and vinegar @ 2 per cent acts as a preservative. Small-sized Aonla fruits, which are not suitable for preparation of preserve or other confectionery items, may be utilized for pickle-making (Goyal *et al.,* 2008). To improve upon the texture of the fruit and also to remove astringency, brining is important in pickling. For preparation of Aonla pickle, it is necessary to add oils, spices, *etc.,* and to leave it for a few days in the sunshine. After a few days, when the pickle is ready, it has to be stored at room temperature.

Sl.No.	*Ingredients*	*Quantity*	*Sl.No.*	*Ingredients*	*Quantity*
1	Aonla pulp	1 kg	4	Fenugreek powder	30 g
2	Salt	125-150 g	5	Fennel powder	50 g
3	Chilli Powder	10 g	6	Turmeric powder	10 mg

Technological Flow-Chart for Processing of Pickle (Aonla)

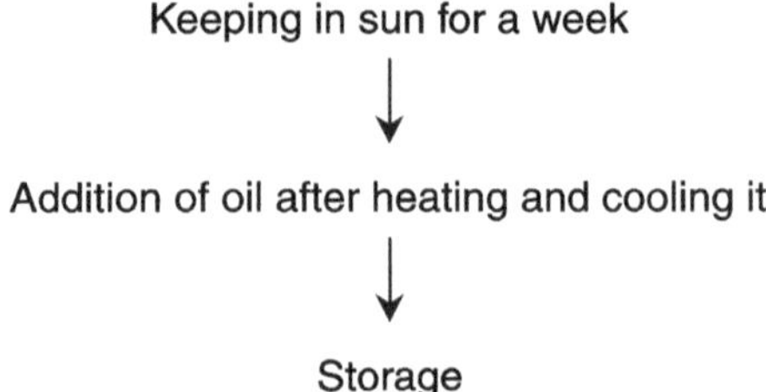

Aonla Shred

This product is very rich in vitamin C content even in dry form. The stability of ascorbic acid in dried product of Aonla is assigned to the presence of leucoanthocyanins or poly-phenols. Processing techniques of Aonla shreds is very simple. Here, we slice fruit into shreds after through washing. Mixed shreds into 4 percent salts and dry up to 15 percent moisture content. Packing and storage is done in 400 gauge polyethylene pouches. Kumar and Sagar (2012) studied on quality Aonla segments. They treated the Aonla segment at different concentration and temperature, and found that Aonla segment was better at 60 ^{0}Brix sugar concentrate at 60 ^{0}C temperature.

Aonla Supari

In the traditional process, Aonla is cut into small pieces and then dried in the sun that darkens the colour of the product besides causing great losses in its ascorbic acid content. Tandon *et al.* (2003b) have reported an improved method in which fresh Aonla fruit is cut into 5-6 pieces and blanched in hot water for 5-6 minutes. Blanched pieces are treated with salt, which extracts water from the fruit pieces by osmosis. After removal of leached, product is dried in tray drier at 60-70 ^{0}C. In this improved method about 50 per cent of ascorbic acid is retained besides improvement in the colour and texture. According to Shekhawat *et al.* (2014) Acidity, moisture content and water activity declined slightly during storage period and retention of ascorbic acid in supari was highest in Banarsi cultivar (42.8 per cent) followed by Chakaiya and NA-7 *i.e.* 41.07 and 40.06 per cent respectively.

Aonla Chavanprash

Chavanprash is a health tonic mentioned in the Indian system of medicine, *i.e.*, Ayurveda. It is prepared by mixing the Aonla pulp and sugar while cooking. In this mixture, spices and medicinal plant extracts are added for further cooking. Chavanprash is claimed to be the best immuno-modulator (Rasayana) formulation, specifically recommended for all metabolic diseases, *viz.*, chronic infections, pulmonary tuberculosis, asthma and cough. It is interesting to note that several of its ingredients still are used ethno medicinally for all aforesaid diseases, thus supporting the properties attributed to it. It has been successfully used as a preventative and curative tonic and is also useful during chronic constipation and urinary infections (Anon, 1952; Gupta *et al.*, 1981). It revitalizes the metabolic functions and cures gas, cough and abnormality (Anon, 1952; Ojha *et al.*, 1973).

Sl.No.	*Ingredients*	*Quantity*	*Sl.No.*	*Ingredients*	*Quantity*
1	Aonla pulp	1 kg	14	Ashwagandha	10 g
2	Sugar/Gud	1.5 kg	15	Satavari	10 g
3	Butter	100 g	16	Bala	5 g
4	Linseed oil	10 g	17	Jeevanti	5 g
5	Javitri	10 g	18	Pushakarmul	5 g
6	Jayphal	10 g	19	Bayaskashta	5 g
7	Clove	10 g	20	Haritaki	5 g
8	Small cardamom	10 g	21	Guruchi	5 g
9	Black pepper	10 g	22	Nilkamal	5 g
10	Large cardamom	10 g	23	Dasmul	4 g
11	Dried ginger	10 g	24	Abharak bhasm	1 g
12	Small pipali	10 g	25	Muktasukti pishti	1 g
13	Banslochan	10 g			

Technological Flow-Chart for Aonla Chawanprash

Selection of fruit (mature)

↓

Washing and sorting

↓

Add 5 litre water in a deep pan and boil it and mix the liquid materials

↓

Add Aonla fruits in muslin cloth and deep in the pan during boiling (low flame)

↓

Remove the muslin cloth when fruit or fruit pieces becomes ripen (loosening of pulp)

↓

Remove the seeds from Aonla fruits and collect the pulp

↓

Boiling will be continue with mix up of liquid materials till the total volume become 1 to 1.5 ltre

↓

Strain it and keep it in one Jar

↓

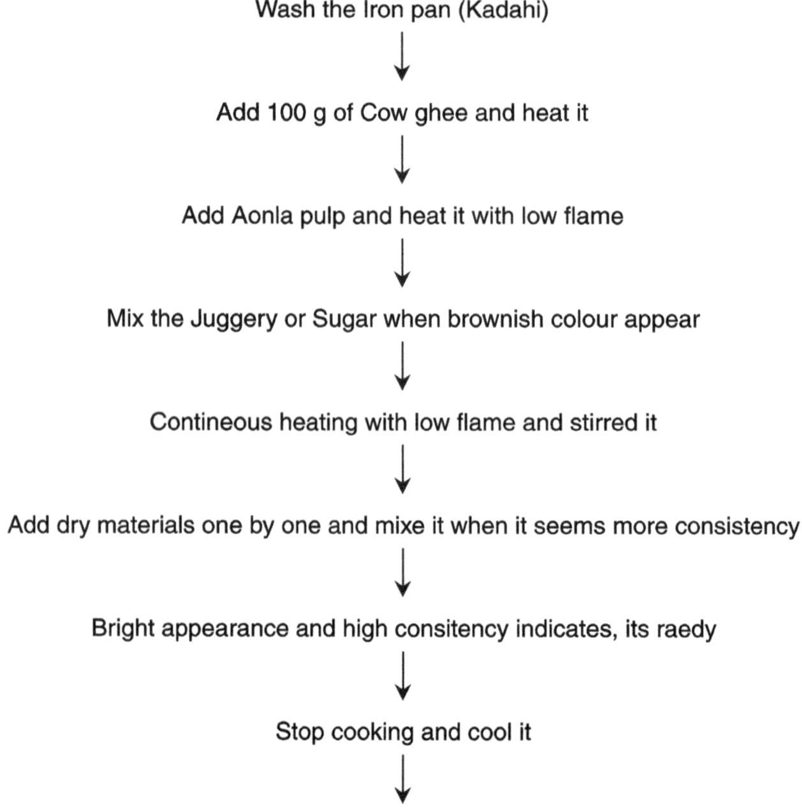

Liquid Materials (Dravya Samagri): Shatawari, Bala, Jeevanti, Pushakarmul, Bayashkasta, Haritki, Guruchi, Nilkamal, Gabharak bhasm, Muktasukti pishti *etc.*

Dry Materials (Sukhi Samagri): Javitri, Jayaphal, Clove, Small Cardamom, Large Cardamom, Black pepper, Dried ginger, Small pipli, Banshlochan, Ashwagandha *etc.*

Aonla Hair Oil

Herbs and herbal drugs are clinically proved good for hair growth. Hair loss problem is of great concern to both males, females and the main problems associated with hair loss are hair fading, dandruff and falling of hair. Various synthetic medicines are available for hair loss which does not treat permanently and also shows severe side effects (Jain *et al.,* 2016). The dried as well as fresh fruits of Emblica officinalis are used for hair oil preparation. The oil is extracted by solvent extraction using petroleum ether. It is the richest source of Vitamin C. It contain minerals like phosphorous, calcium and iron, phyllemblin, tannins and amino acids like glutamic acid, aspartic acid, proline, alanine and lysine. The oil contains linoleic, linolenic, oleic acids and myristic, palmitic and stearic acids (Dhar *et al.,* 1951). The fixed oil obtained from the fruits possess hair growth promoting activity. The dried fruits are boiled in coconut oil, application of this oil prevents graying of hair (Kumar *et al.,* 2012). It can also be used to cleanse the hair, to control dandruff, to prevent hair fall (Sharma and Agarwal 2003).

Triphala

What is Triphala? Simply put, Triphala is an herbal formula used in the ancient Science of Ayurveda. The word 'Triphala' is translated as 'three fruits'. These fruits, also known as 'myrobalan plums' are Amalaki (*Emblica officinalis* Gaertn.), Bibhitaki (*Terminalia bellerica* Roxb.) and Haritaki (*Terminalia chebula* Retz.) (Frawley, 2003, Chauhan *et al.,* 2013). There is a folk saying in India which says, "No Mother? Do not worry so long as you have Triphala". This is allusion to the belief that Triphala cares for internal organs, much in the same way a mother takes care of her children, and among laxative herbs, Triphala is the 'safest and most strengthening' (Frawley, 2003). It can be use in digestion, gouty arthritis, wound healing compounds, anti-oxidant, anti-cancer, free radicle and protection against gamma radiation. Bibhitaki, Haritaki and Amalaki in 1:2::4 formulation contains a higher proportion of antioxidants which would be responsible for its significant effect on hyperlipedimia against triphala 1:1:1 formulation (Chouhan *et al.,* 2013, Mukherjee *et al.,* 2006).

Aonla Churan

Tandon *et al.* (2005) standardized the procedure and recipe for preparation of churan from dried Aonla powder. The retention of Vitamin 'C' is hardly 50 per cent. Therefore, alternately a ecofriendly option through integration of solar tunnel drying can be employed, which can be designed as per requirement and produces high retention capacity as far as Vitamin 'C' is concerned in final products.

Biscuits

Dietary fibre, Vitamin C and antioxidant-enriched biscuits were developed by incorporation of Aonla pomace (a by product generated in Aonla juice processing). The dietary fibre content of the finished product was about 5 times higher than thermal disinfestations set-up for pulses has a batch capacity of 100 kg the control, while Vitamin C and antioxidant concentration were 15.6 mg/100g and 0.25 g, respectively. Biscuits had a shelf-life of more than 3 months when wrapped in 100-gauge polypropylene pouches under ambient conditions. The biscuits prepared in accordance with the invented process can be supplemented as fibre, Vitamin C and antioxidant fortified diet for children and adult alike (Shekhawat *et al.,* 2014). The fibre-enriched biscuits may be helpful in curing the constipation and other ailments related to fast food habits (CISH, Lucknow).

Mouth Freshener

Aonla supari available in the market suffers huge processing losses in Vitamin C and other nutrients. This defeats the purpose of producing the product. Since dehydrated Aonla pulp retains sufficient amount of Vitamin C, there is a need to develop an alternate chewing product with health promoting nutrients. Therefore, develop Aonla mouth freshner, which can prove to be a novel innovative product that can provide a better substitute. Nutritive and palatable mouth freshners were prepared from dehydrated Aonla (*Emblica officinalis* Gaertn) pulp of 'Desi' and 'Banarsi' cultivars by mixing carboxy methyl cellulose, gums, arecanut, cardamom, sugar and milk powder at different proportions as a substitute for pan masala, tobacco and gutka. Mouth fresheners developed were packed in high density polyethylene pouches (HDPE, 100 gauge), stored at ambient conditions (8–20 ^{0}C, 60 per cent RH). During storage for 6 months, ascorbic acid and overall acceptability of mouth freshener decreased ($p \leq 0.05$) and moisture content increased. The equivalent relative humidity of mouth freshener was 49 per cent and 53 per cent in 'Desi' and 'Banarsi' cultivars, respectively. Mouth freshener prepared with 50 per cent dehydrated Aonla pulp, 15 per cent fennel, 10 per cent areca nut and 20 per cent sugar (Shekhawat *et al.,* 2014).

REFERENCES

Agarwal, S. and Chopra, C.S. (2004). Studies on changes in ascorbic acid and total phenols in making Aonla products. *Bev. Food World.* **31**(5): 32–33.

Ahmad, J., Mehmood, Z., Mohammad, F. (1998). Screening of some Indian medicinal plants for their antimicrobial properties. *Journal of Ethanopharmacology.* **62**: 183–193.

Alam, M.S., Amarjit, S. and Sawhney, B.K. (2010). Response surface optimization of osmotic dehydration process for Aonla slices. *J. Food Sci. Tech.* **47** (1): 47–54.

Alam, S. and Singh, A. (2005). Process for dehydration of Aonla powder. 39th Annual Convention of Indian Society of Agricultural Engineers. ANGR Agricultural University, Hyderabad, March 9–11, p. 222.

Anonymous (1952). Raw materials, in: Shastri B.N. (Ed.), Wealth of India, Vol. III, CSIR, New-Delhi, India, 1952.

Asenjo, C. F. (1953). The story of West Indian Cherry (*Malpighia punicifolia* L.). Boletin del Colegio de Quimicos de Puerto Rico. **10**: 8-11.

Asmawi, M. Z., Kankaanranta, M., Moilanen, E., Vapaataslo, N. H. (1993). Anti-inflammatory activities of *Emblica officinalis* Gaertn Leaf extracts. *Journal of Pharmacy and Pharmacology.* **45**: 581–584.

Barat, J. M. E.; Chiralt, A. and Fito, P. (1998). Equilibrium in cellular food osmotic solution system as related to structure. *J Food Sci. Technol.,* **63**: 836-840.

Barthakur, N. N. and Arnold, N. P. (1991). Chemical analysis of Emblica (*Emblica officinalis* Gaertn) and its potential as food source. *Horticultural Science.* **47** (1/2): 99-105.

Chandan K., Prashanth S.J., Nataraj S.K., Indudhara S.M., Rokhade A.K. (2010). Preparation of dehydrated slices and RTS beverages from Aonla (*Emblica officinalis* Gaertn.) fruits, *Int. J. Agric. Sci.* **6** (1): 300–304.

Chauhan, O. P.; Srivastav, S.; Pandey, P. and Rai, G. K. (2005). A study on the development of Aonla blended sauce. *Beverage Food World.* **32** (5): 31-32.

Chouhan, B., Kumawat, R. C., Kotecha, M., Ramamurthy, A. and Nathani, S. (2013). Triphala: A comprehensive aurvedic review. *International Journal of Research in Aurveda and Pharmacy.* **4**(4): 612-617.

Damame, S. V., Gaikwad, R. S., Patil, S. R. and Masalkar, S. D. (2002). Vitamin C content of various Aonla products during storage. *Orissa J. Hort.* **30**: 19–22.

Deka, B. C., Sethi, V., Prasad, R. and Batra, P. K. (2001). Application of mixtures methodology for beverage from mixed fruit juice/pulp. *J Food Sci. Technol.,* **38** (6): 615-618.

Deokar, A. B. (1998). Medicinal Plants Grown at Rajegaon, first ed. DS Manav Vikas Foundation, Pune: 48–49.

Dhar, D. C., Dhar, M. L. and Shrivastava, D. L. (1951). Chemical examination of the seeds of Emblica officinalis Gaertn.: Part I—The fatty oil and its component fatty acids. *Sci Industr Res* 10B: 88–91.

Dhumal, S. S., Karale, A. R., Garande, V. K., Patil, B. T., Masalkar S. D. and Kshirsagar, D. B. (2008). Shelf life of Aonla fruits: Influenced by postharvest treatments and packaging materials. *Indian J. Agri. Res.* **42** (3): 189 -194.

Frei, B., England, L. and Ames, B. (1989). Proceedings of Natural Academic Sciences, USA. **86**: 6377–6381.

Ganachari, A., Thangavel, K. and Eswaran, S. (2005). Postharvest Processing, Potential of Aonla (*Emblica officinalis*) fruit. *Food and Pack,* **5**(5): 34-35.

Geetha, N. S., Kumar, S., Sandooja, J. K. and Rana, G. S. (2006). Effect of osmotic concentration process on biochemical characteristics of Aonla preserve during storage. *Haryana J. of Hort. Sci.,* **35** (3 and 4): 242-244.

Goyal R. K., Patil R. T., Kingsly A. R. P., Himanshu W., Pradeep K. (2008). Status of postharvest technology of Aonla in India – A review. *Am. J. Food Technol.* **3**: 13–23.

Gupta O. P., Srivastava T. N., Gupta S. C., Badola D. P. (1981). Ethnobotanical and phytochemical screening of high attitude plants of Ladakh, *Bull. Medico-Ethno Bot. Res.* **2**: 67–88.

Ihantola-Vormisto, A., Summanen, J., Kankaanranta, H., Vuorela, H., Asmawi, Z. M., Moilanen, E. (1997). Anti-inflammatory activity of extracts from leaves of *Phyllanthus emblica. Planta Medica.,* **63**: 518–524.

Jain S. K., and Khurdiya D. S. (2004). Vitamin C enrichment of fruit juice based ready-to-serve beverages through blending of Indian gooseberry (*Emblica officinalis* Gaertn.) juice. *Plant Foods Hum. Nutr.* **59**: 63–66.

Jain, P. K., Das, D. and Jain, P. (2016). Evaluating hair growth acitivity of herbal hair oil. *International Journal of Pharm Tech Research.* 9(3): 321-327.

Jain, S. K. and Khurdia, D. S. (2009). Ascorbic acid content and non-enzymatic browning in stored Indian gooseberry juice as affected by sulphitation and storage. *J Food Sci. Technol.,* **46** (5): 500-501.

Jain, S. K., Khurdia, D. S., Guar, Y. D. and Lodha, M. L. (2003). Thermal processing of Aonla (*Emblica officinalis* Gaertn) juice. *Indian Food Packer.* **47** (1): 46-49.

Jamwal, K. S., Sharma, I. P. and Chopra, L. (1959). Pharmacological investigations of the fruits of *Emblica officinalis. J. Sci. Ind. Res.* **18**: 180-181.

Jayshri, S. and Jolly, C. I. (1993). Phytochemical antibacterial and pharmacological investigations on *Monordica chiranlia* and *Emblica officinalis. Ind. J. Pharm. Sci.* **1**: 6-13.

Jeena, K. J., Joy, K. L. and Kuttan, R. (1999). Effect of *Emblica officinalis. Phyllanthus amarus* and *Picrorhizia kurroa* on N-Nitrosodiethyl amine induced hepatocarcinogenesis. *Cancer Letter.* **136**: 11–16.

Jose, J. K., Kuttan, Y., Kutan, R. (2001). Antitumour activity of *Emblica officinalis. Journal of Ethanopharmacology.* **75**: 65–69.

Kumar, P. S. and Sagar, V. R. (2012). Effect of concentration and temperature of osmotic solution on mass transfer kinetics and its influence on quality of Aonla (*Emblica officinalis*) segments. *Indian Journal of Agricultural Sciences.* **82**(4): 318-322.

Kumar, S. and Nath, V. (1993). Storage stability of amla fruits: a comparative study of zero-energy cool chamber versus room temperature. *J. Food Sci. Tech.* **30**(3): 202–203.

Kumar, S., Swarankar, V., Sharma, S. *et al.* (2012). Herbal cosmetics: used for skin and hair. Inventi Rapid Cosmeceuticals. (4):1–7.

Mishra, P., Srivastava, V., Verma, D., Chauhan, O. P. and Rai, G. K. (2009). Physico-chemical properties of Chakiya variety of Aonla (*Emblica officinalis*) and effect of different dehydration methods on quality of powder. *African Journal of Food Science.* **3** (10): 303-306.

Mukherjee, P. K., Rai, s., Bhattacharyya, S., Debnath, P. K., Biswas, T. K., Jana, U., Pandit, S., Saha, B. P. and Paul, P. K. (2006). Clinical study of triphala-A well known phytomedicine from India. *Iranian Journal of Pharmacology and Therapeutics.* **5**(1): 51-54.

Nath, V. and Sharma, R. K. (1998). Screening of Aonla (*Emblica officinalis* Gaertn.) cultivars for processing. *Prog. Hort.,* **30** (1 and 2): 76-77.

Nath, V., Singh, I. S. and Kumar, S. (1992). Evaluation of Aonla cultivars for their shelf-life at ambient temperature. *Narendra Deva J. Agri. Res.* **7**(1): 117.

Nayak, P., Tandon, D. K. and Bhatt, D. K. (2012). Study on changes of nutritional and organoleptic quality of flavoured candy prepared from Aonla (*Emblica officinalis* G.) during storage. *International Journal of Nutrition and Metabolism.* **4**(7): 100-106.

Ojha J. K., Bajpai H. S., Sharma P. V., Khanna M. N., Shukla P. K. and Sharma T. N. (1973). Chavanprash as an anabolic agent – Experimental study (Prelim. work), *J. Res. Indian Med.* **8** (2): 11–13.

Pareek S., Rathore N. S. and Kaushik R. A. (2011). Aonla (*Emblica oficinalis* Gaertn.): Postharvest handling and processing technology, Rajasthan Coll. Agric., Techn. Bull. No. 1, Udaipur, India: 45 p.

Patel, A. B. and Sachan, S. C. P. (1995). Extension of storage life of Aonla fruits. *Indian Food Pac.* **49** (4): 43-46.

Pathak, P. K., Divedi, P. and Kumar, S. (2009). Effect of postharvest treatments on shelf life of Aonla (*Emblica officinalis*) fruits damaged during harvesting. *J Food Sci. Technol.,* **46**(3): 283-285.

Perry, L. M. (1980). Medicinal Plants of East and South East Asia: Attributed Properties and Uses. MIT Press, Cambridge.

Pokharkar, S. M., Prasad, S. and Das, H. (1997). A model for osmotic concentration of banana slices. *J Food Sci. Technol.,* **34**: 230-235.

Prajapati, V. K., Nema, P. K. and Rathore, S. S. (2009). Effect of pretreatment and recipe on quality of solar dried Aonla (*Emblica officinalis* Gaertn) shreds. Food security and environmental sustainability. Indian Institute of Technology, Kharagpur, December, 17–19.

Premi, B. R., Sethi, V. and Maini, S. B. (1999). Effect of steeping preservation on the quality of Aonla (*Emblica officinalis* Gaertn.) fruits during storage. *J Food Sci. Technol.,* **36**(3): 244-247.

Rajkumar, N. V., Theres, M. and Kuttan, R. (2001). *Emblica officinalis* fruits afford protection against experimental gastric ulcers in rates. *Pharmceut. Biol.,* **39**(5): 375-380.

Ram R. B., Meena M. L., Sonkar P., Lata R., Upadhyay A. K. (2011). Standardization and evaluation of blended Aonla (*Emblica officinalis* Gaertn.) and bael (*Aegle marmelos* Correa) RTS beverages, *Plant Arch.* **11**(1): 205–208.

Ranote, P. S. and Singh, S. (2006). Value added Aonla products. The Schumacher Centre for Technology and Development Bourton-on-Dunsmore Rugby, Warwickshire, CV23 9QZ United Kingdom. Website: http://practical action. org/practicalanswers/

Rao, T. S., Kumari, K. K., Netaji, B. and Subhokta, P. K. (1985). Ayurveda Siddha. *J. Res.* **6**: 213-224.

Rastogi, R. P. (1993). Compendium of Indian Medicinal Plants, CDRI, Lucknow and ID, New Delhi **1**: 530.

Rege, N. N., Thatte, U. M., Dahanukar, S. A. (1999). Adaptogenic properties of six rasayana herbs used in Ayurvedic medicine. *Phytotherapy Research*. **13**: 275–291.

Sagar, V. R. and Kumar, R. (2006). Preparation and storage study of ready-to-eat dehydrated gooseberry (Aonla) shreds. *J. Food Sci. Tech.* **43**: 349–352.

Sahu, G. D., Singh, P. and Singh, A. K. (2010). Studies on physic-chemical changes in Aonla preserve (Murabba) of three cultivars during storage. *Research Journal of Agricultural Sciences*. **1** (4): 419-425.

Sairam, K., Rao, Ch.V., Dora Babu, M., Vijay Kumar, K., Agrawal, V. K. and Goel, R. K. (2002). Anti-ulcerogenic effect of methanolic extract of *Emblica officinalis*: an experimental study. *Journal of Ethanopharmacology*. **82**: 1–9.

Shankar, G. (1969). Aonla for your daily requirement of vitamin C. *Ind. Hort.*, **13**: 9-15.

Sharma, L. and Agarwal, G.A. (2003). Medicinal plants for skin and hair care. *Indian Journal of Traditional Knowledge*. **2**(1):62–68.

Sharma, S. K., Perianayagam, J. B., Joseph, A., Christina, A. J. M. (2003). Anti-inflammatory activity of ethanol and aqueous extracts of *Emblica officinalis* Gaertn fruits. *Hamdard Medicus* XLVI, 71–73.

Sharma, S. R., Alam, S. and Gupta, S. (2002). Storage study on dehydrated Aonla powder. XXXVI Annual Convention of ISAE, Indian Institute of Technology, Kharagpur, January 28–30. p. 155–156.

Shekhawat, S. Rathore, N. S. and Kaushik, R. A. (2014). Advance in processing and product development of Aonla (*Emblica officinalis*) in Indian context- A review. *International Journal of Food and Nutritional Sciences*. 3(6): 242-247.

Singh, B. P., Pandey, G., Sarolia, D. K., Pandey, M. K. and Pathak, R. K. (2005). Shelf-life evaluation of Aonla cultivars. *Indian J. Hort.* **62**: 137- 140.

Singh, I. S., Pathak, R. K., Dwivedi, R. and Singh, H. K. (1993). Aonla production and postharvest technology. Tech. Bulletin, *Narendra Dev University of Agr. and Tech.* Kumarganj, Faizabad.

Singh, R. and Kumar, S. (1997). Studies on the effect of different storage conditions on decay loss of Aonla *var.* Chakaiya. *Haryana J. Hort. Sci.* **26**: 9-12.

Singh, R. and Kumar, S. (2000). Studies on the effect of postharvest treatments on decay loss and biochemical changes during storage of Aonla (*Emblica officinalis* G.) fruit cv. Chakaiya. *Haryana J. of Hort. Sci.*, **29** (3 and 4): 178-179.

Singh, S., Singh, A. K., Joshi, H. K., Bagle, B. G. and More, T. (2010). Effect of zero energy cool chamber and postharvest treatments on shelf-life of fruits under semi-arid environment of Western India. Part 2. Indian gooseberry fruits. *J Food Sci. Tech.* **47**(4): 450–453.

Singh, V., Singh, H. K. and Singh, I. S. (2004). Evaluation of Aonla varieties (*Emblica officinalis* Gaertn.) for fruit processing. *Haryana J. of Hort. Sci.*, **33** (1 and 2): 18-20.

Sivakumar, V. and Sundaram, K. S. (2010). Effect of different maturity stages on fruit quality traits or physiochemical characters of Aonla *cv.* Na-7. *Plant Archives.* **10**(2): 903-905.

Suresh, K. P. and Sagar, V. R. (2009). Influence of packaging materials and storage temperature on quality of osmo-vac dehydrated Aonla segments. *J Food Sci. Technol.*, **46**(3): 259-262.

Tandon, D. K., Dikshit, A. and Kumar, S. (2005). Development of churan from dried Aonla powder. *Beverage and food world.* **32**: 53-54.

Tandon, D. K., Kumar, S. and Diskhit, A. (2003b). Improvement in technology for Aonla supari. *Processed Food Industry.* **6**(8): 23.

Tandon, D. K., Yadav, R. C., Sood, S., Kumar, S. and Dikshit, A. (2003a). Effect of blanching and lye peeling on the quality of Aonla peeling. *Indian Food Packer.* **57**(6): 147-152.

Thakur, C. P. and Mondal, K. (1984). Effect of *Emblica officinalis* on cholestral induced atherosclerosis in rabbits. *Ind. J. Med. Res.*, **79**: 142-146.

Tripathi, V. K., Singh, M. B. and Singh, S. (1988). Studies on comparative compositional changes in different preserved products of amla (*Emblica officinalis* Gaertn) var. Banarasi. *Indian Food Packer.* **42**(4): 60-66.

Verma, R. C. and Gupta, A. (2004). Effect of pre-treatments on quality of solar-dried amla. *J. Food Eng.* **65**: 397–402.

Vijayanand, P., Kulkarni, S. G., Reena, P., Aksha, M. and Ramana, K. V. R. (2007). Effect of processing on gooseberry fruits and quality changes in dehydrated gooseberry powder during storage. *J. Food Sci. Tech.* **44**: 591–594.

Vinayagamoothy, T. (1982). Antibacterial activity of some medicinal plants of Srilanka Ceylon. *J. Sci. Bilo.*, **11**: 50-55.

Yadav, V. K. and Singh, H. K. (1999). Studies on the preharvest application of chemicals on shelf-life of Aonla (*Emblica officinalis* Gaertn.) fruits at ambient temperature. *J. Appl. Hort.* **1**(2): 118-121.

2

Bael

Scientific Name: *Aegle marmelos* Corr.

Family: Rutaceae

Edible Portion: Succulent placenta

Bael (*Aegle marmelos* Corr.) is an indigenous fruit of India belongs to family Rutaceae, which is generally grown in Southeast Asia and has various nutritional and therapeutic properties. It is commonly known as Bengal quince (John and Stevenson, 1979), Indian quince, Golden apple, Holy fruit, Bel, Belwa, Sriphal, Stone apple and Maredo in India. Bael fruit is a sub-tropical, deciduous tree and fruit is globuse with grey or yellowish hard woody shell. The history of this tree has been traced to Vedic period 2000 BC – 800 BC. The Bael fruit has been mentioned in Yajurveda, which was introduced into Europe from India in 1959 (John and Stevenson, 1979). The tree has great mythological significance and it abounds in the vicinity of temples. Inside this, there is soft yellow or orange coloured mucilaginous pulp with numerous seeds, which are densely covered with fibrous hairs and are embedded in a thick, gluey, aromatic pulp (Kaushik *et al.,* 2002). The pulp of fruit contains many functional and bioactive compounds such as carotenoids, phenolics, alkaloids, coumarins, flavonoids, terpenoids, and other antioxidants, which may protect against chronic diseases (Singh *et al.,* 2015a). Bael fruit is truly popular for its ability to combat constipation. Its medicinal properties have been described in the ancient medical treatise in Sanskrit in *Charaka Samhita* (Aiyer, 1956). All parts of this tree stem, bark, root, leaves and fruit at all stages of maturity have medicinal virtues and have been used as medicine for a long time. The unripe Bael fruits are used for pharmaceutical use (Hema and Lalitha Kumari, 1999; Pattanayak and Mohapatra, 2008), therapeutic use and preparation of jams, marmalade and syrups (ITDG, 2000). The fruit's medicinal value is very high when it just begins to ripen. The fruit is aromatic, cooling and laxative. It is useful in preventing or curing scurvy (Kartikar and Basu, 1935) and it also strengthens the stomach and promotes its action. Bael is considered to be one of the richest source of riboflavin (Mukharjee and Ahmad, 1957) and it provides lots of minerals and vitamins to the diet (Barthakur and Arnolds, 1989). The pulp also contains a balsam-like substance, and 2-furocoumarins-psoralen and marmelosin ($C_{13}H_{12}O_3$), highest in the pulp of the large, cultivated forms (Singh and Nath, 2004). Another interesting point about Bael is the fact that it is equally useful while still unripe, whereas other fruits become consumable on ripening. Bael fruit comes under the underutilized fruits, because it's neither grown commercially on large scale nor traded widely. It is cultivated, traded and consumed locally (Singh, 2007).

Importance

Key to giving a fillip to India's food production would be minimization of all the wastages in the food distribution. Wild varieties of plants, yielding edible fruits growing throughout the Himalayas, contributed directly to cultural heritage of India. Even today, these fruits are eaten in plenty by local people, as they are commonly available in abundance in their habitats. Bael fruit have many advantages in terms of cultivation, hardy in nature and production of good crop even under adverse conditions. These underutilized fruit crops play crucial role in successful running of processing industry round the year in the region. A large proportion of rural population depends on locally available fruits to meet their dietary requirements. These fruit crops have their own history of consumption, local people are well aware their nutritional and medicinal properties (Patel *et al.,* 2008).

Nutritive Value/Benefits

Bael fruit is highly nutritive with a great medicinal use and the richest source of riboflavin. Gehlot and Dhawan (2005) reported about all parts of the trees *viz.* root, bark, leaves, flowers or fruits are used for curing one or other human ailment. Bael is one of the deciduous sacred trees of the Hindus associated with Gods having useful medicinal properties, especially as a cooling agent. Leaves are offered in prayer to Shiva and Parvathi since ancient times (Rajasekaran and Meignanam, 2008). The fruit pulp has detergent action and has been used for washing clothes. The shell of hard fruits has been fashioned into pill- and snuffboxes, sometimes decorated with gold and silver. Cologne is obtained by distillation from the flower (Patkar *et al.,* 2012). The roots are sweet, astringent, bitter and febrifuge. They are useful in curing dyspepsia, dysentery, diarrhea, vitiated condition of vata, vomiting, cardiopalmus, stomachalgia, intermittent fever, seminal weakness, swelling, uropathy and gastric irritability in infants. The bark decoction for malaria and leaves are useful in opthalmia, deafness, and diabetes and asthmatic complaints. The flowers allay thirst vomiting. The unripe fruits are acrid, astringent, bitter, digestive, sour, stomachia and are useful in dysentery-diarrhea and stomachalgia. The ripe fruits are sweet, aromatic, cooling, febrifuge, laxative, good tonic for heart and brain and cure dyspepsia (Parichha, 2004; Chowdhury *et al.,* 2008; Mitra 1999).

Bael has high content of tannin which makes its an effective cure for dysentery and cholera. There is as much as 9 per cent tannin in the pulp of wild fruits, less in the cultivated types and rind contains up to 20 per cent tannin content. The different parts of Bael are used for various therapeutic purposes, such as for treatment of asthma, anemia, fractures, healing of wounds, swollen joints, high blood pressure, jaundice, diarrhoea healthy mind and brain typhoid troubles during pregnancy (Sharma *et al.,* 2011).

Chemical Composition of Fresh Bael Fruit (Food Value/100 g edible portions)

Nutrients	*Quantity*	*Nutrients*	*Quantity*
Edible portion	64 per cent	Calcium	85.00 mg
Water	61.5 g	Iron	0.60 mg
Protein	1.80 g	Thiamine	0.13 mg
Fat	0.39 g	Riboflavin	0.03 mg
Fiber	2.90 g	Niacin	1.1 mg
Minerals	1.70 g	Ascorbic acid	65-100 mg
Carbohydrates	31.80 g	Acidity	0.25-0.36 per cent
Carotene	55.00 mg	Viscosity	12.7-19.0 per cent

Source: Meena ***et al.,*** 2017.

Harvesting and Yield

In India flowering occurs in April to May, soon after the new leaves appear and the fruit ripen in 10 to 11 months from bloom March to June of the flowering year (Orwa, 2009; Chopra and Nayer, 1956). Fruits are ready for harvest in North India in April-May when the fruits shell changes its colour from deep green to yellowish green. The fruit should be picked individually, so that it does not fall on the ground to avoid cracking of the shell, which may lead to spoilage during storage. The stem end of fruit is prone to infection. Therefore the fruits should be harvested along with a portion of stalk (approximate 2cm). The stalk automatically separated from the fruit when the fruit is ripe, thus it is an indication of ripening. Care should be taken to avoid dropping of Bael fruits on the ground during harvest and handling as it results in cracking or even internal injury due to impact which leads to rotting of fruits during storage and ripening. To avoid fruit falling on the ground fruit picker is used for harvesting. It is necessary that fruits to be graded based on size, packed with some cushioning materials (paper cuttings or paddy straw @ 2 per cent of total fruit weight) to avoid cracking of the fruits due to impact in transportation (Meena *et al.,* 2017). Bearing in budded/grafted plants starts 3-4 years after planting, whereas in seedling trees 7-8 years. A 12-15 years old Bael tree produces 300-500 fruits and the individual fruit was found to vary between 500g to 3000g, which varies greatly in Bael varieties.

Mature Bael fruits ripe in 2-3 weeks under ambient conditions, if treated with ethrel solution @ 5 ml/liter in water for 20 minutes, ripe in 1-2 weeks with proper colour, flavor and quality development. Ethrel solution once prepared can be used 3-4 times for fruit treatment. Large size fruits take longer time in ripening as compared to small sized fruits. Premature Bael fruits harvested before April could also be ripened with ethrel treatment (Anon. 2011).

1. Handling of Bael Fruits after Harvest

- ☆ Care is needed when handling Bael fruits to avoid causing cracks in the rind as this can lead to fungal infections.
- ☆ Remove immature and injured fruits.
- ☆ Grade the remaining fruits according to their size. There is no standard practice for the grading of Bael fruits.
- ☆ Wash fruits using chlorinated water (100 ppm) and drain them.

2. Packaging and Storing of Fresh Fruits

- ☆ Fruits are packed in gunny bags, baskets or wooden crates for transportation, storage and marketing. Use cushioning material, such as straw or paper, when packing Bael fruits.
- ☆ Fruits harvested at full maturity (light green colour) can be kept for about 15 days at 30 ^{0}C.
- ☆ Fruits harvested ripe can be stored for only one week at 30 ^{0}C.
- ☆ Ripe fruits can be stored for about 3 months at 9 ^{0}C and relative humidity of 85-90 per cent. After that moulds is likely to develop at the stem end and at any cracks in the rind.

3. Pre-processing into Pulp

- ☆ Break fruits using a strong knife or special Bael breaking equipment.
- ☆ Scoop out the pulp along with the seeds and fibres and discard the peel.
- ☆ Add water equal to weight of pulp (1:1).
- ☆ Add 5 g citric acid per kg pulp.
- ☆ Mix the pulp.
- ☆ Heat mixture to 80 ^{0}C for one minute.
- ☆ Pass mixture through a pulping machine or stainless steel sieve of 20 meshes. Discard seeds and fibers.
- ☆ Treat pulp with 1000 ppm of SO_2 (optional). Dissolve sodium metabisulphite (1.5 g/kg) in little water and add.
- ☆ Place pulp in airtight containers and seal.
- ☆ Store pulp for up to 6 months for further processing

Physical Characteristics

The Bael fruits are hard-shelled berry usually globule (6-12 cm diameter) with a nearly smooth pericarp. Pericarp of Bael fruit is 3 mm thick, hard and filled with soft and yellow fragrant pulp. The raw fruit, due to its high acidic nature and astringent taste, is unacceptable to consumers. Seeds are many, compressed and arranged in closely packed tiers in the cell surrounded by very tenacious, slimy, transparent mucilage, which becomes hard when dried. The testa is white with wooly hairs and the embryo has large cotyledons (Hume, 1957; Singh and Roy, 1984). Some variety like NB-9 are good in taste (35-40 ^{0}Brix) and average fruit yield is 70-80 kg and rind of fruit is thin, flesh contains less fibre and seed, whereas CISH-B-1 and CISH-B-2 are good in taste (35-41 ^{0}Brix), thin rind (0.1-0.12 cm) and average yield is about 40-60 kg (Peter, K.V. 2007).

Chemical Composition

The chemical composition of Bael fruits is influenced by environmental factors. The chemical composition of fresh fruit in respect of water, carbohydrate, protein, fat, minerals, carotene, thiamine, riboflavin, niacin and vitamin C has been reported by lot of researchers (Meena *et al.,* 2017). The fruits are rich in carbohydrate, protein and riboflavin. Jauhari *et al.* (1969) reported that the Bael fruit, TSS and acidity ranged from 1283 to 2818 g, 28 to 36 per cent and 0.256 to 0.368 per cent, respectively, and according to Jauhari and Singh, 1971, Bael fruit contains 28-39 per cent TSS, 19-21 per cent carbohydrates, 11-17 per cent sugar, 1 per cent protein, 0.2 per cent fat, 7-21 mg/100g vitamin C. In addition, it is rich in vitamin A (186 IU/100gm pulp); volatile oils and marmelosines. Its food value is 88 calories/100gm. Thus, it is richer than most of the reputed fruits like apple, guava and mango, which have a calorific value of only 64, 59 and 36, respectively.

Storage

The storage life of the fruit depends upon the stage of harvesting. Bael fruit can be stored for 10-15 days at normal temperature, whereas fruits harvested at ripe stage can be stored for a week. Ripe Bael fruits can be stored for 2 weeks at 27-32 ^{0}C. Fruits below 9 ^{0}C temperature are greatly injured and brown spots appear on the fruit surface (Ravani and Joshi, 2014). The storage life of Bael fruit could be increased from 2 weeks at 30 ^{0}C to 12 weeks at 9 ^{0}C and 85-90 per cent relative humidity (Roy and Singh, 1979). The ripe Bael fruit could be made available 2-3 months prior to schedule with the treatment ethrel (1000-1500 ppm) and storing the fruits at 30 ^{0}C after harvesting in January. It took 18-24 days for the fruit to be artificially ripened. The composition of fruits, ripened artificially or naturally, did not vary much, with slightly less sugar content reported in artificially ripened fruits (Roy and Singh, 1981). Bhadra and Sen (1998) reported that fruits wrapped with blue cellophane paper, butter paper, perforated polyethylene bagging + $KMnO_4$ and perforated polyethylene bagging, which showed maximum extension of shelf life of 24 days over other treatments. Among the chemicals by which fruits were treated, NAA 100 ppm, GA_3 50 ppm, stannous chloride 250 ppm and cobalt nitrate 250 ppm were effective in extension of storage life and marketability of fruits. Singh (1998) mention that NAA and calcium nitrate were effective in minimizing the loss in weight, reduced the rate of respiration, spoilage percentage and finally maintained the edible quality and marketability of fruits during storage of guava.

Value Added Products and their Uses

A large number of Bael-processed products (Preserve, candy, panjiri, toffee, jam *etc.*) are prepared and some scientist and researcher are already worked on their processed products (Singh and Nath, 2004; Singh *et al.*, 2013). Rakesh *et al.*, 2005 standardized the recipe of Bael processed products. Morton (1987) explained the medicinal composition and uses of Bael fruit. The fresh ripe pulp of the higher quality cultivars, and the "sherbet" made from it are taken for their mild laxative, tonic and digestive effects. A decoction of the unripe fruit, with fennel and ginger, is prescribed in cases of hemorrhoids. It is employed in the treatment of leucoderma.

Bael Products and their Recipe

Products		*Preserve*	*Candy*	*Panjiri*	*Toffee*	*Jam*
Recipe	Fruit pulp (kg)	1	1	1	1	1
	Sugar (kg.)	1.25	1.25	1.5	0.5	0.75
	Citric acid (g)	2-3	2-3	-	-	3-4
	Water (ltr.)	1	1	-	-	-
	Desi-ghee (kg)	-	-	1	-	-
	Roasted wheat flour	-	-	Q.S.	-	-
	Dry fruits	-	-	Q.S.	-	-
	Glucose	-	-	-	100	-
	Skimmed milk powder (g)	-	-	-	150	-
	Butter (g)	-	-	-	100	-
Remark		-	-	Prescribed for Stomach ailments	-	Minimum 55 per cent of fruit portion and minimum 68 per cent of TSS

Q.S.: Quantity sufficient.

Source: Rakesh *et al.*, 2005.

Preserve and Candy

Preserve and candy are prepared from mature (tender green fruit), whole or large pieces of fruits in which sugar is impregnated till it becomes tender and transparent minimum fruit portion and minimum total soluble solids in preserves should be 55 and 70 per cent, respectively (Lal *et al.*, 1960). Fruits in general contain more than 75 per cent water and get spoiled quickly if not stored properly. Removal of water from fruits is known to help in longer period of storage. The osmotic dehydration techniques not only enables the storage of fruits for a longer period

Technological Flow-Chart for Processing of Preserve and Candy

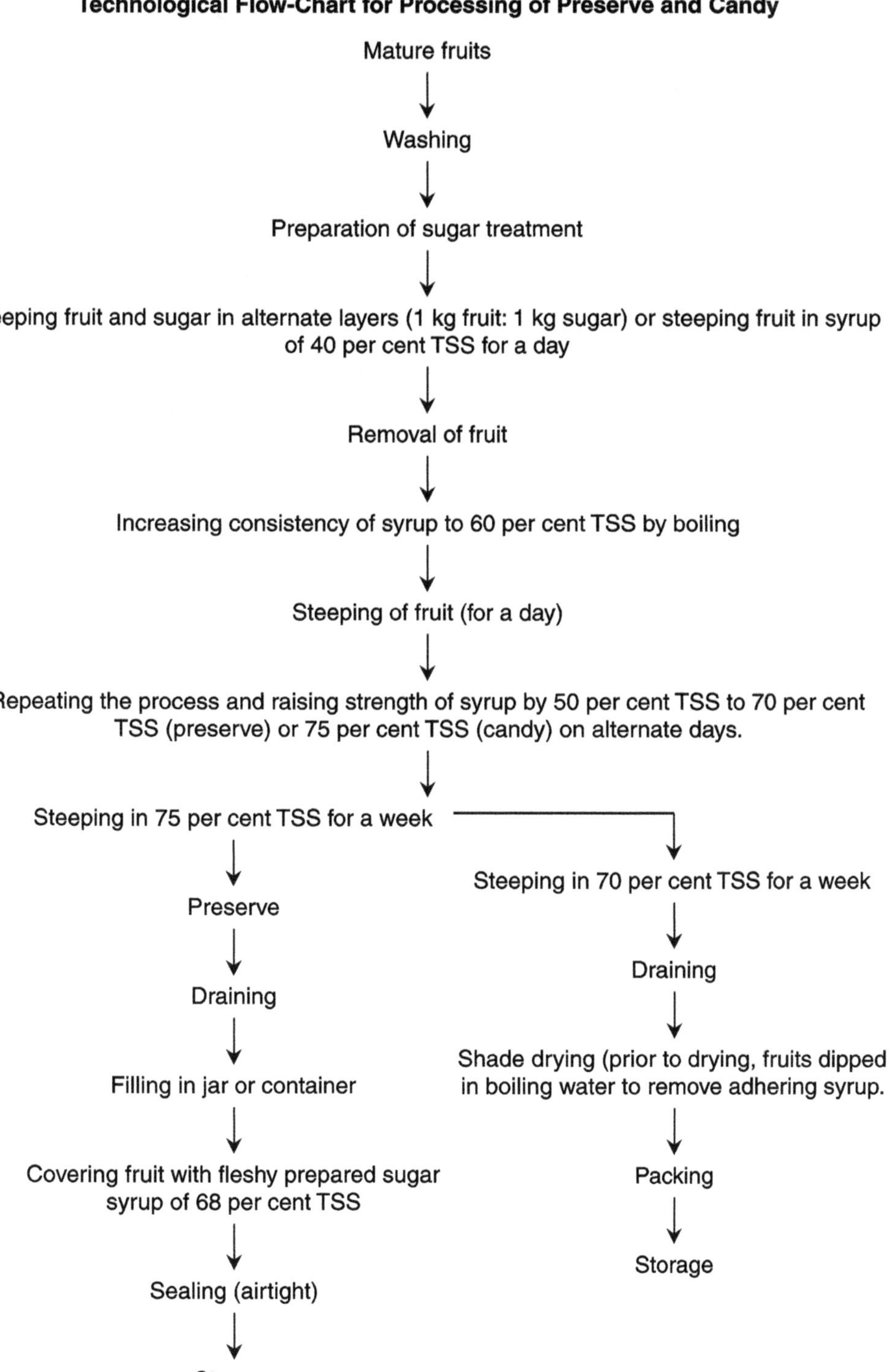

Source: ITDC (2000).

but also preserve the flavor, colour and texture of the product to a great extent and prevents its microbial spoilage (Bongirwar, 1997).

"A fruit of its pieces impregnated with cane sugar or glucose syrup, subsequently drained free of syrup and dried is known as candied fruit". The total sugar content of the impregnated fruit is kept at about 75 per cent to prevent fermentation. In case of Bael candy, the fruit slices are drained subsequently free of syrup and dried at 55-60 ^{0}C for 8-10 hrs in oven. According to Singh *et al.* (2015a) Bael preserve was prepared with combination of sugar, acid, water and preservative, and two cultivars of Bael like Local cultivar of West Bengal and NB-5 were evaluated at monthly interval of their quality and found that cv. NB-5 was safer than Local cv. upto 12 months. According to Singh *et al.* (2017a) Bael fruit pieces were dipped in alum and lime solution @ 2 per cent for two hrs and then blanched it at 8 kg/ cm^2 for 28 min. before impregnated with sugar syrup and dying the fruit pieces, gave the best result in respect of quality and storage behaviors up to 3-8 months in safer condition. The similar finding was also reported by Singh *et al.* (2017) in Bael candy storage at room and refrigerated storage.

Bael Fruit Squash

An ideal composition of Bael fruit squash was found to be 50 per cent extracted pulp, 50 ^{0}Brix and 1 per cent acidity. The squash was chemically preserve by addition of 300 ppm SO_2 (Roy and Singh, 1979). Fruit beverages commercially contains at least 25 per cent fruit pulp or juice and 40-50 per cent TSS, besides 1 per cent acid (Srivastav and Kumar, 1993). The squash from Bael fruit pulp was prepared by adjusting the TSS and by adding the preservatives like sodium metabisulphite @ 350 ppm SO_2 (Bhat and Kaul, 2006), and sodium benzoate @ 1g/litre (Verma and Gehlot, 2006). The squash was then filled in sterilized bottles, crowned and pasteurized at 80 ^{0}C for 30 minute fallowed by cooling and wax sealing to insure air tightness (Kenghe, 2008). Fresh juice is bitter and pungent fruit extract lower the blood sugar (Kirtikar and Basu 1995; Dhankhar, 2010).

Technological Flow-Chart of Processing of Bael Fruit Squash

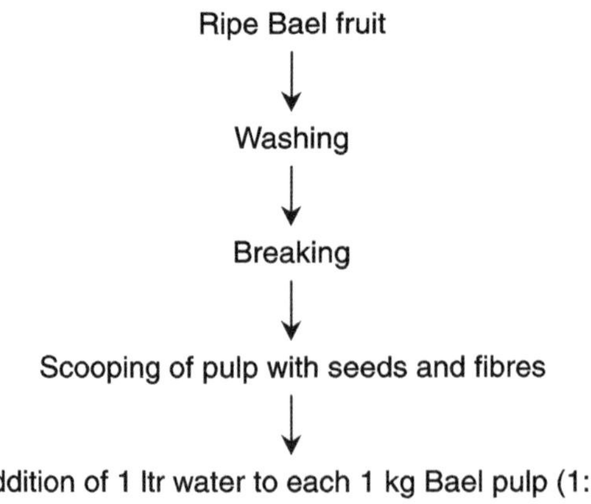

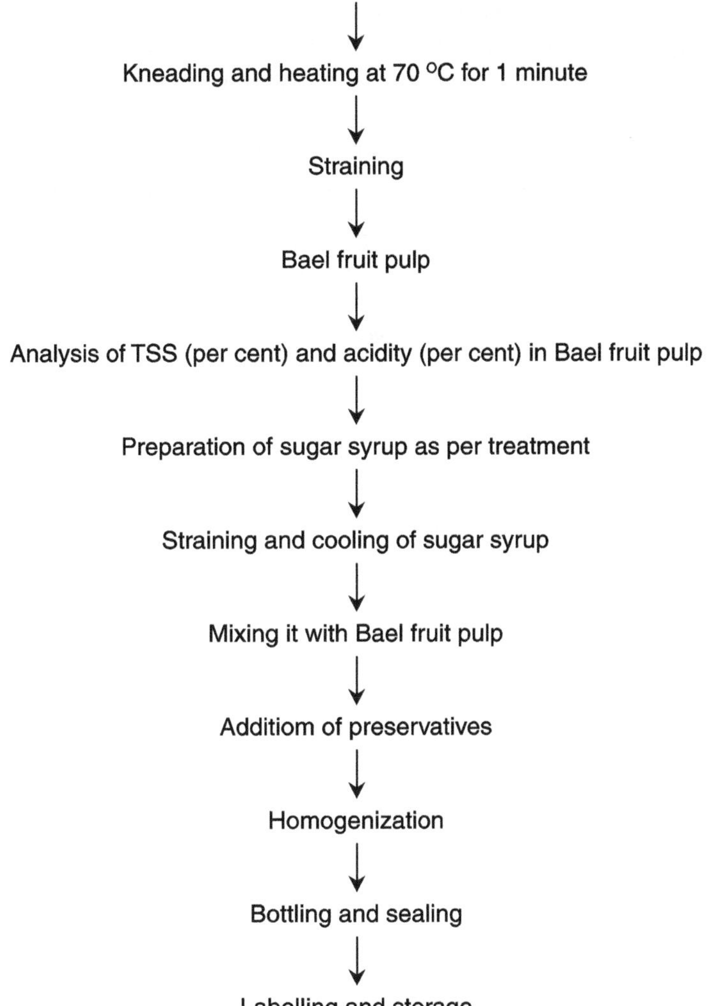

Bael Fruit Pulp

The ripe fruit were washed with tap water and broken by striking against hard object. The fruit pulp along with its seeds and fibres was scooped with the help of stainless steel spoon. Amount of water equal to the weight of pulp was added. The mixture of pulp and water was then heated up to 80 °C for 1 minute and cooled. Pulp free from seeds and fibres was then obtained by passing through 20-mesh stainless steel sieve. The extracted Bael pulp was improved by adjusting the TSS by addition of sugar and acidity by the addition of citric acid (Chand and Gehlot, 2006). Fruit pulp can be stored for 6 months, when stored in heat-sealed containers (Ravani and Joshi, 2014). The Bael fruit pulp contains many functional and bioactive compounds such as carotenoids, phenolics, alkaloids, coumarins, flavonoids, and terpenoids and has innumerable traditional medicinal uses (Karunanayake *et al.*, 1984; Singh 1986; Nagaraju and Rao, 1990). Fruit pulp marmalade is used as prevention during cholera epidemics, also given to prevent the growth of piles; useful in patients suffering from chronic dysenteric condition relieves flatulent colic from a condition of chronic gastrointestinal eatarrh (Patkar *et al.*, 2012).

Technological Flow-Chart of Processing of Bael Fruit Pulp

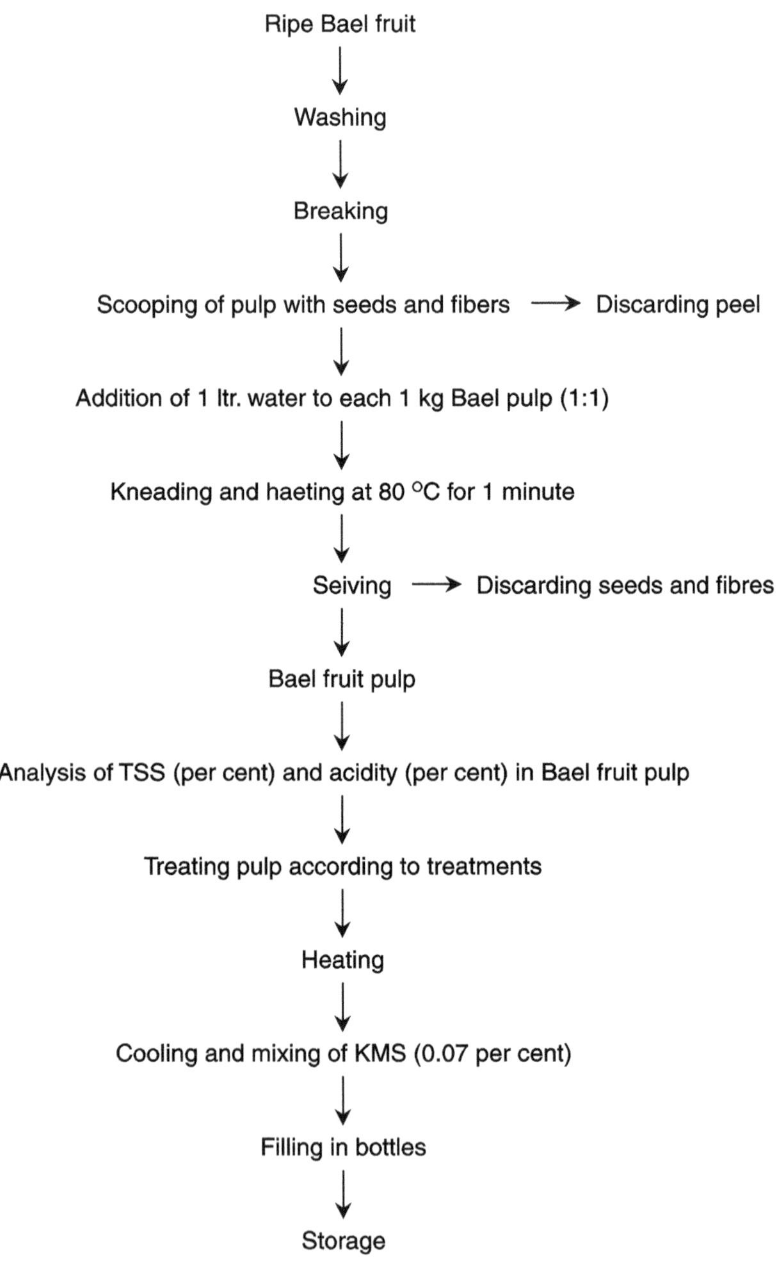

Dehydrated Bael

Select mature green fruits, wash and cut 1-1.5 cm thick slices of fruit pulp after removing its hard shell. Fumigate these slices of fruits with sulphur dioxide fumes for an hour in sulphur box and dehydrated at 55-60 °C in oven up to a constant weight. Pack dried slices in polyethylene bags or glass jar for future use by Rakesh *et al.,* 2005.

Technological Flow-Chart of Processing of Dehydrated Fruit (Bael)

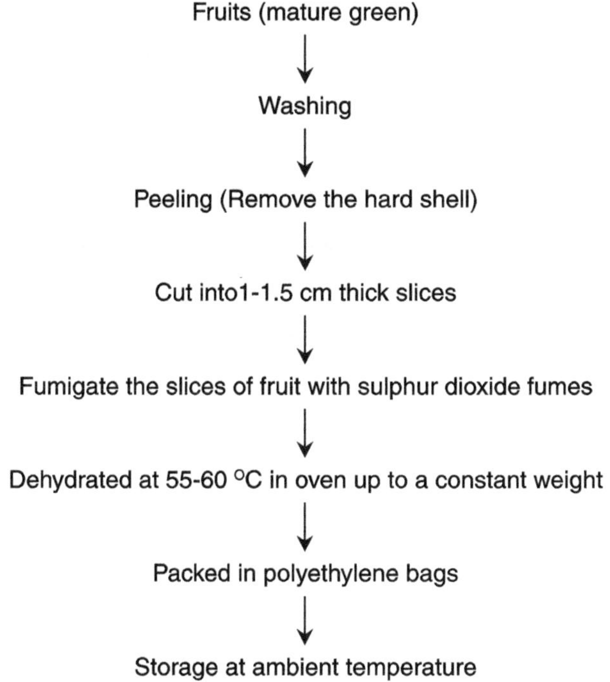

Bael Ready-To-Serve (RTS)

For preparing Bael ready-to-serve (RTS) drink TSS (per cent) and acidity (per cent) in the extracted Bael fruit pulp were analyzed and on the basis of analysis, requisite amounts of sugar and citric acid dissolved in required amount of water were added to measure quantity of Bael pulp for adjustment of TSS (per cent) and acidity (per cent) in RTS drink as per treatments. The prepared RTS drink was thoroughly homogenized and filled in clean, sterilized 200 ml capacity glass bottles leaving 2-3 cm headspace. The beverage filled bottles were sealed with sterilized crown corks, pasteurized in boiling water for half an hour, cooled in air and stored in cool and dry place (Verma and Gehlot, 2006 and 2007). According to Rathod *et al.* (2014) described the preparation of RTS from Bael and aonla mixed pulp with the best combination was 2:3 which was pasteurized by thermal heat at 90 °C was more inactivating the microbial flora. However, the shelf life of juice was established within 45 days.

Technological Flow-Chart of Processing of Bael Ready-to-Serve

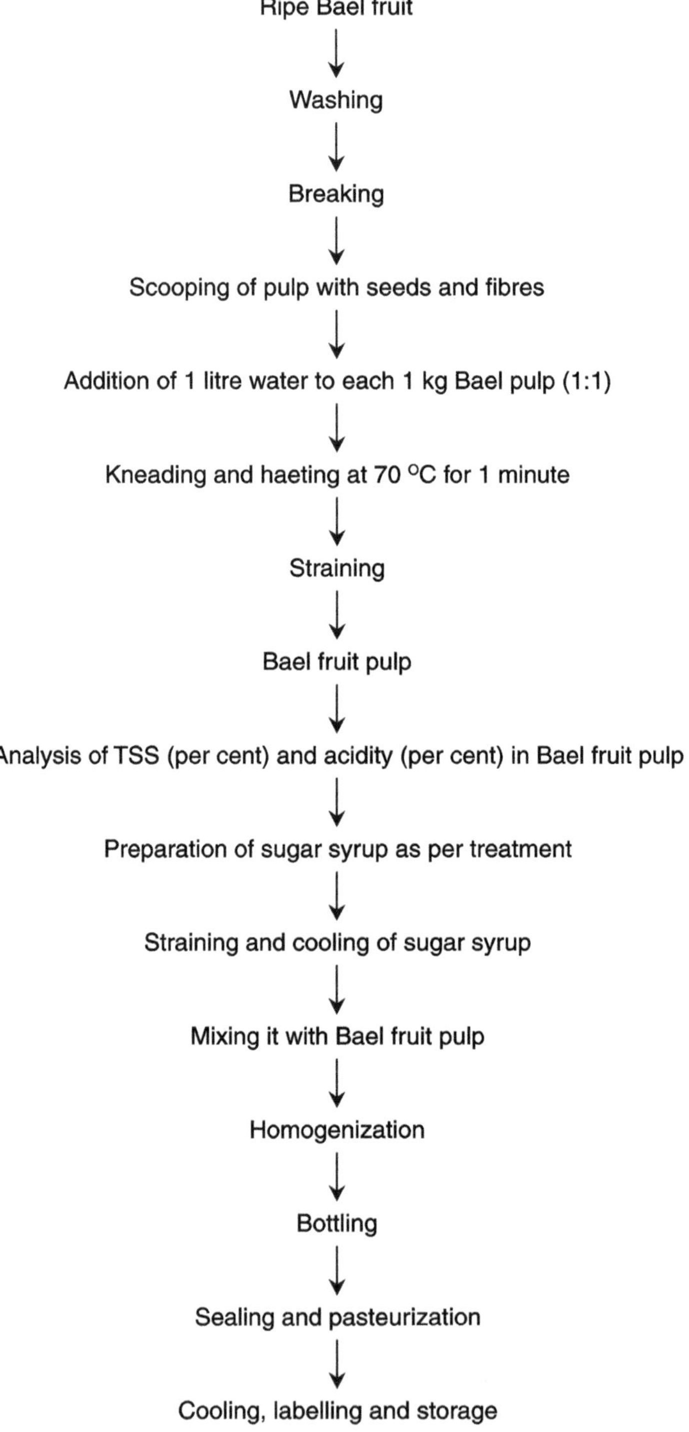

Bael Powder

Bael powder can be prepared by simply grinding fruit slices in a grinder. Pack ground bale powder in polyethylene bags and store in dry places after proper sealing for consumption in future. According to Rani *et al.* (2009) the colour of Bael pulp powder was found golden poppy in direct sun drying, light brownish in hot water, orange in hot sand and orange/burnt orange and rust in oven. Before this, fruits were cracked and treated with A) hot water at 70 °C for 1 hour, B) hot sand for 2 hours, C) oven for 2 hours at 70 °C, after that, rind of Bael fruit was broken and removed the Bael pulp, and dried in direct sun light. Roy and Singh, 1979 reported that Bael fruit powder was prepared by drying the pulp to a thin sheet to below 4 per cent moisture level and then grinding to powder. Fruit powder can be stored for a year when packed in 400 gauge polypropylene pouches and stored under dark, cool place, while fruit jam, squash and preserve can be stored for several months (ITDC, 2000). According to Singh *et al.* (2015b) five air drying temperatures (60, 65, 70, 75 and 80 °C) with five thickness of pulp on the tray (2, 4, 6, 8 and 10 mm) used for Bael fruit powder and obtain the powder prepared from the pulp dried at 65 °C with a drying thickness of 2 mm was optimum with respect to drying time, colour and ascorbic acid content.

Technological Flow-Chart of Processing of Fruit Powder

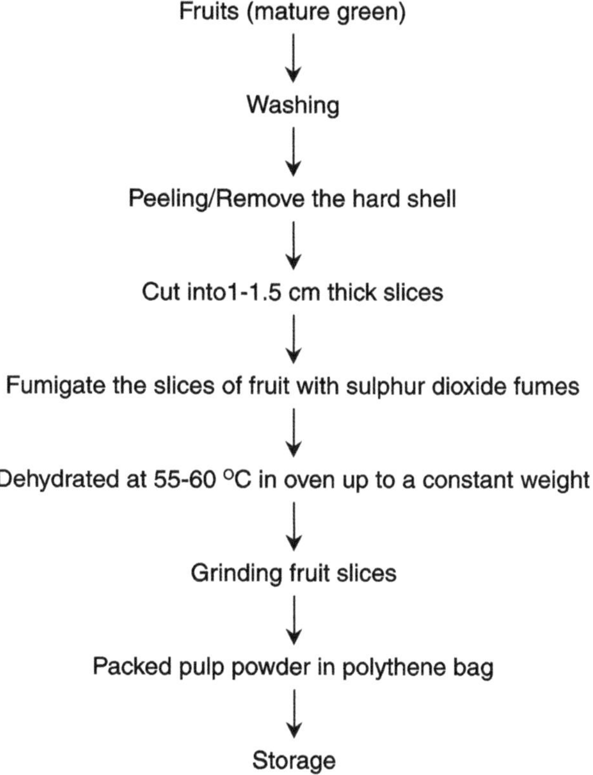

Jam

The Bael fruit can be processed to prepare jam. A mix fruit jam from Bael and mango was prepared by Mishra and Chopra (2006). Jam is a concentrated fruit product processing a fairly heavy body, reach in natural fruit flavour. Pectin in fruit gives it a good set and high concentration of sugar facilitates its preservation. It is

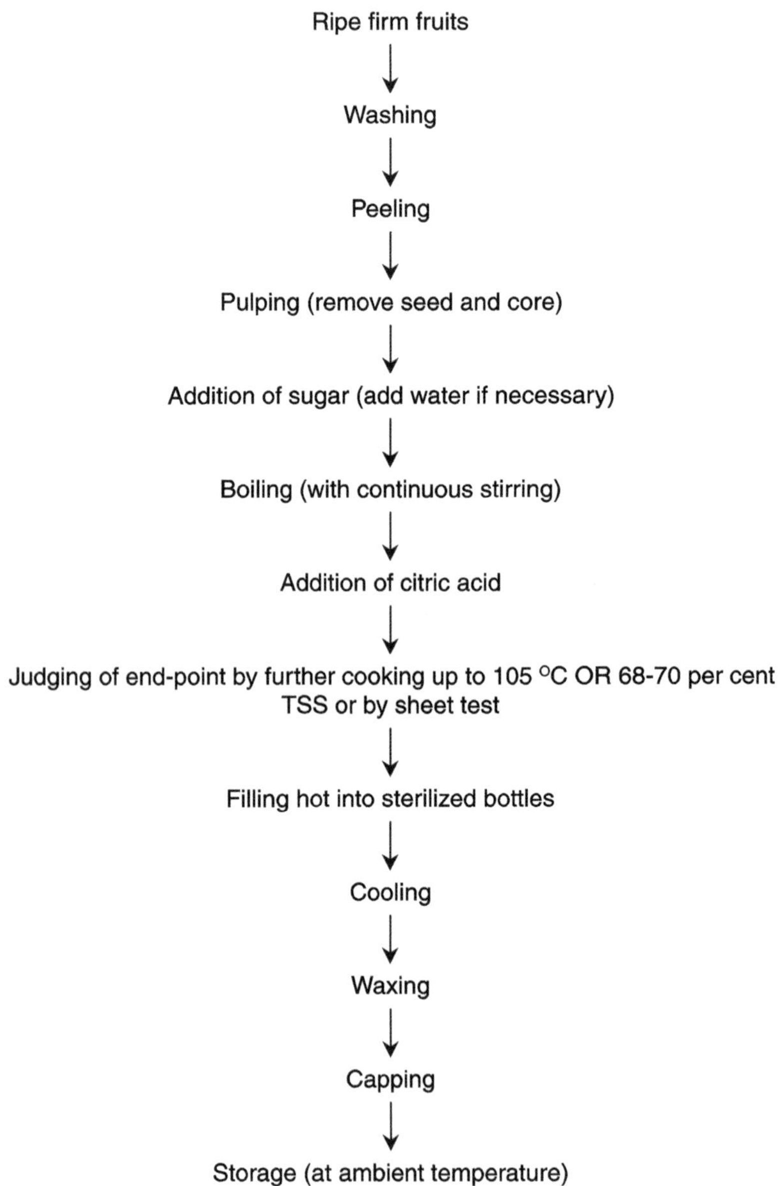

prepared by boiling the fruit pulp and juice with sufficient quantity of sugar to a reasonably thick consistency to hold the fruit tissues in position. A fruit jam should contain minimum 45 per cent of fruit portion and minimum 68 per cent of total soluble solids. The methods of making Bael jam are given by Singh and Chaurasiya, 2014. Apart from exploring possibilities to prepare standardized drugs by using different plant parts of *Aegle marmelos*, production of jam by using its fruits should be promoted as health tonic at commercial scale (Patkar *et al.*, 2012).

Slab

It is also known as leather or paper. Ripe fruits are used in its preparation. Wash ripe fruits and collect fruit pulp by breaking fruits and removing its hard shell. At 200-300 ml of water for each one kg of fruit pulp, mix well and heat it up to 80 °C. Collect fruit pulp free of seeds and fibers by straining heated mass through stainless steel sieve. Add sugar, citric acid and potassium meta-bisulphite (KMS) to this pulp so that treated pulp contains 35 per cent total soluble solids, 0.5 per cent total acidity and 0.07 per cent KMS. Boil treated pulp and spread on aluminum trays smeared with butter. Dry at 55-60 °C for 15-16 hrs. to a moisture content of 14.5 per cent. Cut slaves of dried pulp in aluminum trays, wrap in butter paper and pack in polyethylene bags. Addition of up to 10 per cent sugar to the extracted pulp was found to be ideal before drying the pulp to a moisture content of 14.5 per cent (Roy and Singh, 1979; Rakesh *et al.*, 2005).

Technological Flow-Chart of Processing of Fruit Slab

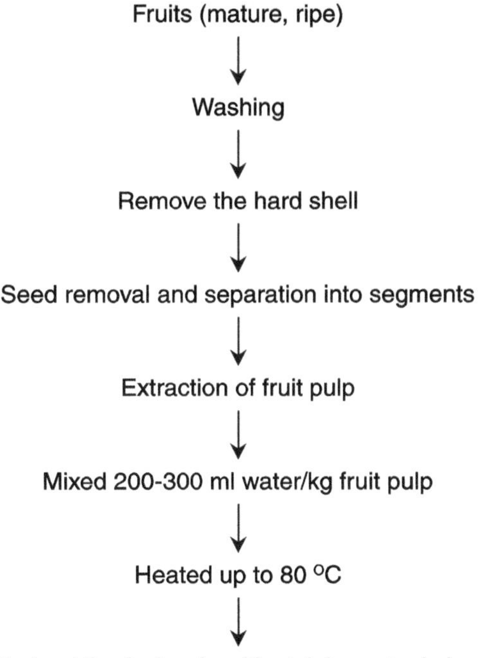

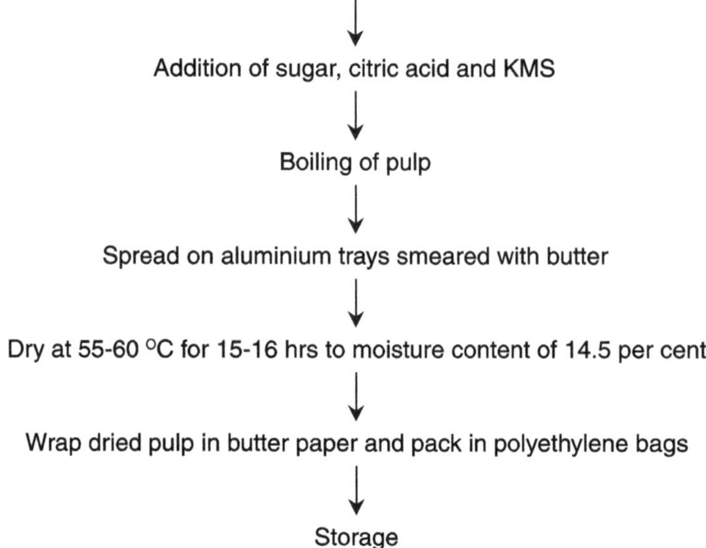

Toffee

Fruit toffees generally are more nutritious than ordinary toffees, and Bael fruit pulp will provide even better toffees because of its nutritional and medicinal properties. Bael fruit toffees was successfully prepared by mixing 40 parts of cane sugar, 4.5 parts of glucose, 10 parts of skim milk powder and 6 parts of hydrogenated fat to 100 parts of extracted pulp. The final moisture content of the toffee was kept at 8.5 per cent (Roy and Singh, 1979). The Bael fruit toffees were prepared from Bael pulp with sugar, ghee, milk powder, corn flour, citric acid and different proportion of cardamom and cinnamon (0.5 g, 1.0 g and 1.5 g respectively). Toffee prepared with 0.5g cardamom was found the best in the case of color and appearance, texture, overall acceptability and 1.0 g cinnamon powder in case of taste and flavor (Bhatt and Verma, 2016).

Technological Flow-Chart for Preparation of Fruit Toffee

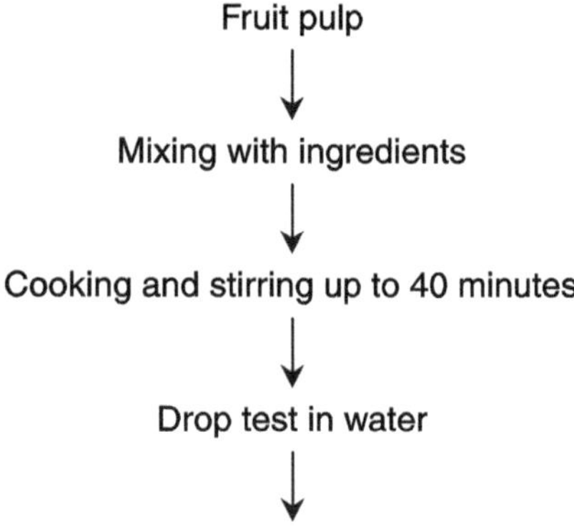

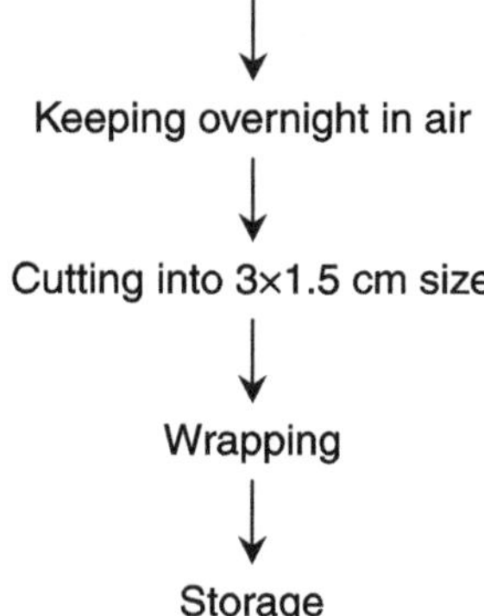

Panjiri

This product is highly nutritive, restorative and is prescribed for stomach ailments. Rakesh *et al.* (2005) used Bael powder 1 kg, desi ghee (butter oil) 1 kg, sugar powder 1.5 kg, wheat flour and dry fruits as per taste for preparation of Bael panjiri. Bael powder is roasted in desi ghee. Roast wheat flour was added in small amount as per taste in desi ghee. All the ingredients were then mixed.

Bael and Aloe Vera Blended Beverage

Mixed the Bael juice and aloe vera juice then added sugar, citric acid and water for preparing strained syrup then added the preservative that is sodium benzoate at the rate of 100 ppm. Syrup with sodium benzoate (preservative) pasteurized in water bath at 60 °C for 5 minutes. After that cold the pasteurized syrup at room temperature and took the bottle of 200 ml capacity filled with pasteurized syrup and then stored it (Mishra and Srivastava, 2013). According to Rathod *et al.* (2014) Bael fruit juice and Indian gooseberry or Aonla juice were optimized to a blended beverage in different ratio, and stored for 45 days in pet bottles (200 ml capacity) at refrigerated temperature. The sample (40:60 ratio) of (Bael: Aonla) indicated the commercial scope for manufacturing good and nutritious.

The Recommended Daily Allowances (RDAs) for energy is regularly exceeded in all developed countries; a major consequence being an increased incidence in obesity, diabetes and associated circulatory problems. Nutritionists have warned against the adverse effects of high intakes of fat, salt, and diets high in sugar and low in dietary fibre (Booth, 1997). Bael fruit is very rich in vitamins, amino acids and minerals when compared to other fruits (Amilasithyapa, 1998), and it can contribute significantly to the daily nutrient needs of the individual. In addition, it can be used advantageously to supplement deficiencies of other foods (Wills *et al.,* 1984). People have a preference for value added products than for consuming it as a completely edible fruit. Preparation of a Bael product with the combination of potatoes, sugar and salt was a new concept. Using a number of ingredients and chemicals appeared unnecessary. The simplicity of the procedure for making such a product will lead to reduce cost of production and increased involvement of small-scale holders in this sector. Products that involve Bael fruit have high medicinal value and can be a better choice for consumers. In other words, this will promote the utilization of Bael fruit in every season.

REFERENCES

Aiyer, A. K. Y. N. (1956). The antiquity of some field and forest flora of India. Bangalore Printing and Publishing Co Ltd, Bangalore.

Amilasithyapa, P. (1998). Rasa Guna Piri Beli. Tharangi Prints, Hilevel Road, Nawinna, Maharagarna.l-Z, l3-18 pp.

Anonymous (2011). Ripe mature Bael fruits with ethrel. Central Institute of subtropical horticulture, Lucknow. Press release on 19th April 2011.

Barthakur, N. N. and Arnolds, N. P. (1989). Certain organic and inorganic constituents in Bael (*Aegle marmelos* Correa) fruits. *Tropical Agriculture*. 3:10-14.

Bhadra, S. and Sen, S. K. (1998). Extension of Storage Life of Bael (*Aegle marmelos* Correa.) Var. Kalyani Selection-1. *Environment and Ecology*. 16(3): 699-703.

Bhat, A. and Kaul, R. K. (2006). Utilization of Bael for Processing. Journal of Research, SKUAST-J, 5(2): 203-208.

Bhatt, D. K. and Verma, S. (2016). A study on development of herbal food product Bael (*Aegle marmelos*) fruit Toffee. *IOSR Journal of Environmental Science, Toxicology and Food Technology*. 10(3): 5-14.

Bongirwar, D. R. (1997). Application of osmotic dehydration for preservation of fruits. *Indian Food Packer*. 51(1): 18-21.

Booth, R. G. (1997). Snack Food. CBS publishers and distributors, Darya Ganj, New Delhi. 17-19

Chand, T. and Gehlot, R. (2006). Utilization of Bael (*Aegle marmelos* Correa.) for preparation and preservation of pulp. Research on Crops. 7(3): 887-890.

Chopra, R. N. and Nayer, S. L. (1956). Glossary of Indian Medicinal Plants. Publication. New Delhi: 8-13.

Chowdhury, M. G. F., Islam, M. N., Islam, M. S., Islam, A. T. and Husain, M. S. (2008). Study on preparation and shelf life of mixed juice based on wood apple and papaya. *Journal of Soil and Nature*. 2(3): 50-60.

Dhankhar, S. (2010). *Aegle marmelos* (Linn.) Correa: A source of phyto-medicine. *Journal of Medicinal Plants Research*. 5(9): 1497-1507.

Gehlot, R. and Dhawan, S. S. (2005). Bael A Valuable Tree. *Food and Pack*. 5(7): 28-29.

Hema, C. G. and Lalitha Kumari, K. (1999). Screening of pharmacological actions of *Aegle marmelos*. *Indian Journal of Pharmacology*. 20: 80-85.

Hume, H. H. (1957). Citrus Fruits. The Macmillan Company 1957.

ITDC (2000). Processing of wild Bael fruit for rural employment and income generation. *ITDC Food Chain*, 27: 15-17.

Jauhari, O. S. and Singh, R. D. (1971). Bael a valuable fruit. *Indian Horticulture.* 16(1): 9-10.

Jauhari, O. S., Singh, R. D. and Awasthi, R. K. (1969). Survey of some important varieties of Bael (*Aegle mormelos* Correa.). *Punjab Horticulture Journal.* 9: 48-53.

John, L. and Stevenson, V. (1979). The complete book of fruit. Angus and Robertson Publishers Sydney.

Kartikar, K. R. and Basu, B. D. (1995). Indian Medicinal Plants, 2nd edn. 1: 499.

Karunanayake, E. H, Welihinda, J., Sirimanne, S. R. and Sinnadorai, G. (1984). Oral hypoglycaemic activity of some medicinal plants of Srilanka. *Journal of Ethno-pharmacology,* 11(2): 223-231.

Kaushik, R. A., Yamdagni, R. and Sharma, J. R. (2002). Changes in quality parameters during processing and storage of processed Bael (*Aegle marmelos*) fruit. *Indian Food Packer.* Jan.-Feb.: 71-76.

Kenghe, R. N. (2008). Bael fruit processing for value addition and employment generation. *Food Pack Com.* 2(8): 10-12.

Kirtikar, K. R. and Basu, B. D. (1995). Indian medical plant. International Book Publication. 1: 499-502.

Lal, G., Siddappa, G. S. and Tondon, G. L. (1960). Preservation of fruits and vegetables. Indian Council of Agricultural Research, New Delhi.

Meena, H. R. Meena, M. S., Kala, S. and Singh, R. K. (2017). Bael: the most natural medicinal fruit of waste-lands. *Popular Kheti.* 5(1): 53-57.

Mishra, H. and Chopra, C. S. (2006). Processing and storage studies on Bael (*Aegle marmelos* Correa) fruit products: Crush and Jam. *Beverage and Food World Journal.* 44-45.

Mishra, M. and Srivastava, A. (2013). To study the physicochemical properties of Bael and Aloe Vera blended beverage. International *Journal of science and Research.* 4(9): 1-4.

Mitra, S. K. (1999). Other Fruits- Bael. Tropical Horticulture. Bose, T. K. (Ed.). Naya Prakashan, Bidhan Sarani Calcutta 700006 India, ISBN81-85421-32-3, 1999; Vol. 1.

Morton, J. (1987). Bael Fruit, In.: Fruits of Warm climates. Julia F. Morton, Miami, Fl.187-190.

Mukharjee, B. and Ahmad. K. (1957). Riboflavin. *Pakistan Journal of Biological and Agricultural Sciences.* 4: 47-51.

Nagaraju, N. and Rao, K. N. (1990). A survey of plant crude drugs of Rayalaseema, Andhra Pradesh India. *Journal of Ethno-pharmacology.* 29 (2): 137-158.

Orwa, C. (2009). An Introduction to *Aegle marmelos. Agro Data.* 4(2): 1-5.

Parichha, S. (2004). Bael (*Aegle Marmelos*): Nature's Most Natural Medicinal Fruit". Orissa Review. pp. 16-17.

Patel, R. K., Singh, A., Yadav, D. S. and Des, L. C. (2008). Underutilized Fruits of North Eastern Region, India. Underutilized and Underexploited Horticultural Crops. K. V. Peter (Ed.). New India Publishing Agency, New Delhi (India). 4: 223-238.

Patkar, A. N., Desai, N. V., Ranage, A. A. and Kalekar, K. S. (2012). A review on Aegle marmelos: A potential medicinal tree. *International Research Journal of Pharmacy*. 3(8): 86-91.

Pattanayak, P. and Mohapatra, P. (2008). Phytopharmacology of *Aegle marmelos* (L.). *Pharmacognosy Reviews*. 2(4): 50-56.

Peter, K. V. (2007). Underutilized and underexploited horticultural crops. Vol. 2, New India Publishing Agency, New Delhi.

Rajasekaran, C. and Meignanam, E. (2008). In vitro evaluation of antibacterial activity of phytochemical extracts from leaves of *Aegle marmelos* Corr. (Rutaceae). *Ethnobotanical Leaflets*. 2: 1124-1128.

Rakesh, Dhawan, S. S. and Arya, S. S. (2005). Product Development: Processed Products of Bael. Processed Food Industry. 8(10): 25-27.

Rani, A., Chawhaan, P. H. and Rathore, M. (2009). Availability, physical characteristics, harvesting and processing of Bael (*Aegle marmelos*) fruits in Chattisgarh forest. *Green Farming*. 2(7): 465-468

Rathod, A. S., Shakya, B. R. and Ade, K. D. (2014). Studies on effect of thermal processing on preparation of Bael fruit RTS blended with Aonla. *International Journal of Research in Engineering and Advanced Technology*. 2(3): 1-6.

Ravani, A. and Joshi, D. C. (2014). Processing for value addition of underutilized fruit crops. *Trends in Postharvest Technology*. 2(2): 15-21.

Roy, S. K. and Singh, R. N. (1979). Bael fruit (*Aegle marmelos*) a potential fruit for processing. *Economic Botany*. 33(2): 203-212.

Roy, S. K. and Singh, R. N. (1981). Studies on ripening of Bael fruits (*Aegle marmelos*). *Punjab Horticulture Journal*. 21(122): 74-82.

Sharma, G. N., Dubey, S. K., Sharma, P. and Sati, N. (2011). Medicinal values of Bael (*Aegle marmelos*) (L.) Corr.: A Review. *International Journal of Current Pharmaceutical Review and Research*. 1(3): 1-11.

Singh, A. K. and Chaurasiya, A. K. (2014). Postharvest management and value addition of Bael (*Aegle marmelos* Corr.). *International Journal of Interdisciplinary and Multidisciplinary Studies*. 1(9): 65-77.

Singh, A. K. and Nath, N. (2004). Development and evaluation of whey protein enriched Bael (*Aegle marmelos*) fruit beverage. *Journal of Food Science and Technology*. 41(4): 432-436.

Singh, A. K., Chaurasiya, A. K. and Chakraborty, I. (2015a). Quality retention in alum treated Bael (*Aegle marmelos* Corr.) preserve. *The Bioscan*. 10 (1): 135-139.

Singh, A. K., Chaurasiya, A. K. and Chakraborty, I. (2017). Quality Improvement of Bael (*Aegle marmelos* Corr.) candy and storage studies. *Journal of Horticulture and Plant Science*. 1(1): 25-29.

Singh, A. K., Chaurasiya, A. K. and Mitra S. (2017a). Quality Attributes of Bael (*Aegle marmelos* Corr.) preserve and candy. *International Food Research Journal*. 24(1): 298-303.

Singh, A., Sharma, H. K., Kumar, N., Upadhyay, A. and Mishra, K. P. (2015b). Thin layer hot air drying of Bael (*Aegle marmelos* Corr.) fruit pulp. *International Food Research Journal*. 22(1): 398-406.

Singh, A., Sharma, H. K., Kumar, S., Upadhyay, A. and Mishra, K. P. (2013). Comparative effect of crude and commercial enzyme on the juice recovery from Bael fruit (*Aegle marmelos* Correa) using principal component analysis. *International Journal of Food Science*, doi:10.1155/2013/239839.

Singh, D. R. (2007). Underutilized Fruits Crops of Andaman and Nicobar Islands. Underutilized and Underexploited Horticultural Crops. K. V. Peter (Ed.). New India Publishing Agency New Delhi (India). 2: 11-54.

Singh, G. (1998). Effect of calcium nitrate and plant growth regulators on the storage of guava. *Indian Journal of Horticulture*. 45: 45-50.

Singh, R. N. and Roy, S. K. (1984). The Bael. ICAR, New Delhi: 25.

Singh, Y. N. (1986). Traditional medicine in Fiji: some herbal folk cures used by Fiji Indians. *Journal of Ethno-pharmacology*. 15(1): 57-88.

Srivastav, R. P. and Kumar, S. (1993). Fruit and vegetable preservation (Principle and practices). International Book Pub. Com. Charbagh, Lucknow, India.

Verma, S. and Gehlot, R. (2006). Development and Evaluation of Bael beverages. *Haryana Journal of Horticultural Sciences*. 35(3 and 4): 245-248.

Verma, S. and Gehlot, R. (2007). Studies on development and evaluation of ready-to-serve (RTS) drink from Bael (*Aegle marmelos* Correa.). *Res on Crops*. 8(3): 745-748.

Wills, R. B. H., Lim, J. S. K., and Greenfield, H. (1984). Composition of Australian Foods. 22. Tomato. *Food Technology in Australia*, 36 (2); 78-80.

3

Ber

Scientific Name: *Ziziphus* spp.

Family: Rhamnaceae

Edible Portion: Epicarp and Mesocarp

Ber is a tropical and subtropical fruit native to the northern hemisphere (Lyrene, 19790). It belongs to the genus *Ziziphus* of the family Rhamnaceae and order Rhamnales. The family has 50 genera and more than 600 species (Bailey, 1947) of which the species *Z. jujube* Mill (Chinese jujube or Chinese date), *Z. mauritiana* Lamk (Indian jujube or Ber) and *Z. spinachristi* (L.) wild (Christ's thorn) are the most important in terms of distribution and economic significance. *Z. mauritiana* is commonly cultivated throughout the northwest of India and in the drier parts of South India (Brandis, 1906; Sebastian and Bhandari, 1990). The name *Ziziphus* is related to an Arabic word used along the North African coast, zizoufo used for *Z. lotus* (L.) Desf., but also related to the ancient Persian words zizfum or zizafun; and ancient Greeks used the word ziziphon for the jujube. Ber is also found in disturbed areas near settlements and along roads in African countries (Johnston, 1972) where fruits are harvested from naturally seeded plantations and sold in local markets. *Z. jujube* Mill is cultivated in China, Korea and in parts of Southeast Africa (Lin and Cheng, 1995). Ber is being cultivated on an estimated area of 22,000 hectares. The yield potential varies from one to two quintals per tree per annum (Yamadagni, 1985). It is considered an underutilized fruit crop in semi-arid regions of the world and can be successfully cultivated in the marginal ecosystem of the subtropics and tropics (Pareek, 2001).

Ber fruit is generally eaten fresh and is a rich source of ascorbic acid, essential minerals and carbohydrates (Jawanda *et al.,* 1980; Pareek *et al.,* 2002). It is richer than apple and mango in vitamin C, protein and minerals and contains higher

Characteristics	*Major Cultivars of Ber*							
	Katha	*Bagwari*	*Umran*	*Chhuhara*	*Ilaichi*	*Karaka*	*Mundia Murhra*	*Narama*
Appearance	Golden yellow	Yellow to reddish brown	Golden yellow	Greenish yellow	Yellow to reddish brown	Yellow to reddish brown	Greenish yellow	Light green
Avg. Fruit wt. (g)	18.5	16	21	12.5	3.6	23	22	17.5
Pulp: stone ratio	25	13.3	19.6	11.3	24.8	-	16.7	12.9
Moisture (per cent)	74.33	77.92	77.81	76.4	73.97	86.6	88.13	79.3
TSS °Brix	21.4	18	22.7	17.2	24.7	11.8	13	16.8
Acidity (per cent)	0.1	0.11	0.29	0.35	0.22	0.31	0.22	0.25
Reducing sugar (per cent)	4.54	5.94	4.38	3.72	3.91	5.88	4	3.7
Total sugar (per cent)	19.65	16.17	14.84	16.23	16.98	9.97	11.1	14.9
Ascorbic acid (mg/100g)	97.76	126.46	150.99	146.69	129.41	103.5	174.6	146.5

phosphorus and iron than orange (Khera and Singh, 1976). In general, the fruits contain 85.9 per cent moisture, 12.8 per cent carbohydrates, 0.8 per cent protein, 0.1 per cent fats, 0.8 per cent iron, 0.03 per cent each of calcium and phosphorus, and 70 I.U. vitamins A/100 g with an energy value of 55 calories/100g (Yamdagni, 1985).

Harvesting

Harvesting of fruits at proper stage of maturity is very important from quality point of view of ripened fruits. Fruits harvested at correct stage of maturity ripen properly giving characteristics colour, flavor and taste. Immature fruits, lack of sweetness and have an acrid taste. Over-mature or fully ripe fruits turn from yellow or golden-yellow colour to red or dark brown. Under-ripe Ber fruits when picked do not ripen properly and never attain the desirable quality. Hence the market value of such fruits decreases considerably. Over ripe fruits, on the other hand, loose their attractive colour and become soft. Such fruits have less storage life and deteriorate in quality. According to Bal *et al.* (1995) pre-harvest spray of ethephon at a concentration of 300 ppm to Ber trees induced uniform ripening of fruits, and the fruits harvested at optimum maturity could be stored for up to forty days at a temperature of 0 to 3.3 ^{0}C and relative humidity of 85 to 90 per cent. Bal and Chohan (1981) reported that the specific gravity of Umran fruits at maturity was less than one. The fruits develop golden yellow colour at maturity. In Delhi *cv.* Gola take only 150 days whereas, *cv.*, Tikadi required 173 days to mature (Raja, 1985). Increase in starch and acidity was observed only during the growth and development of fruits, which declined with the advancement of fruit development (Bhatia and Gupta, 1985). Fruits of the *cv.*, Kaithli should be harvested at the green-yellow stage and the optimum stage for harvesting fruits of the Gola cultivar is the green mature, green-yellow or yellow stage (Siddiqui and Gupta, 1989). Fully ripe fruits (red brown in colour) lose their crisp texture, but are suitable for dehydration. Spraying Ber fruits 10 days before harvest with $CaCl_2$ (1.7 g/L) with 1 per cent Teepol as a surfactant, reduced PLW, delayed colour development and maintained good quality Ber fruits during the storage period (Gupta *et al.*, 1987). Spraying of 1 per cent $CaNO_3$ at the colour turning stage improved the shelf life of fruits at room temperature (Siddiqui *et al.*, 1989; Siddiqui and Gupta, 1989a; Gupta *et al.*, 1989). Applying a pre-harvest spray of ethephon at 300 ppm to Ber trees induced uniform ripening of fruits and the fruits harvested at optimum maturity could be stored for up to 40 days at 0–3.3 ^{0}C and 85–90 per cent relative humidity (Bal *et al.*, 1995a).

Harvesting Season

The harvesting season varies according to cultivar and location. Some varieties ripen as early as October, others ripen from mid-February until mid-March and others in March or mid-March until the end of April. In any location, diverse cultivars can be harvested over a period of months *e.g.* in North India from February to April, in west India from December to January, but in south India in November. In the Asyut Governorate there are two crops a year, the main being in early spring and the second in the autumn but in India there are two or three pickings with a one season, which are done by hand ladders. One worker is capable of manually harvesting about 50 kg of fruit per day. After wrapping in white cloth, the fruits are put into paper-lined burlap bags of 50 kg capacity for carriage to markets throughout the country.

Picking

All the fruit on the tree does not ripen at one time and therefore 4 to 5 or more pickings have to be done in the season. The fruit picking is done by hand using a ladder. Sometimes, the tree is shaken or the branches are beaten with a bamboo stick so that the ripe fruits fall on the ground. Under-ripe bel' fruits do not ripen after picking. Such fruits are acrid and do not have satisfactory sweetness and fetch a very low price in the market. Overripe fruits, on the other hand, lose their attractive colour and become red, loosing their crisp and juicy texture and become slimy. Such fruits are poor in taste and deteriorate fast. It is, therefore, essential to harvest the Ber fruits at a proper stage of maturity. The best index for the correct picking stage is the characteristic maturity colour and softness of a particular cultivar after the fruit has attained full size. By this stage, the fruit would also have its desired sugar-acid ratio.

Grading

Remove under-ripe, over-ripe, damaged and misshapen fruits. Grade the remaining fruits manually or by passing them through sieves into 2 or 3 levels on the basis of size and colour or according to the grading standard which are given below. Wash graded fruits using with chlorinated water (100 ppm) and drain them for further processing or pack them for storing.

Grade	*Standard*
A	Shining yellow, large (>35mm) to medium size (25-35 mm) fruits of uniform shape with no blemishes.
B	Uneven yellow or yellow red, large (>35mm) to medium size (25-35 mm) fruits, of uniform shape with some blemishes.
C	Red, large (>35 mm) to small (<25 mm) fruits. Uneven yellow, small (<25 mm) fruits.

Pack Ber fruits either for proper storage or for safe transport to local or distant markets:

- ✰ In small packages of 1-2 kg in perforated 150 gauge polythene or nylon-net bags or cardboard cartons.
- ✰ In large packages of 10-20 kg in gunny bags, net bags, cloth packages or wooden or plywood boxes with holes or slits.
- ✰ For transportation, corrugated cartons of about 10 kg are the most suitable packaging material. For short distances, cheaper materials can be used if cushioning and ventilation is provided.

- ☆ Ber fruits should be packed in non- organic materials to avoid spoilage caused by microbes during storage and transport.

Packing

The Ber fruits are generally packed in white cloth sheets or in gunny bags. For long distance transportation, the fruits are commonly packed in gunny bags. Sometimes a paper or white cloth lining is kept on the inside of the gunny bag. Jain and Chitkara (1978) packed the fruit of 'Umran' cultivar in gunny bags, in perforated wooden boxes, in wooden boxes with slits, in mulberry baskets and in card-board cartons and observed that after a 800 km travel, the fruit packed in mulberry baskets had minimum shriveling and weight loss and the highest organoleptic rating.

Yield

According to Bakshi and Singh (1974) the average yield from 10-20 year old trees of different Ber cultivars varies between 80-200 kg/tree. Sixteen year old kaithki tree produced a year of 120 kg and Mundia Murhara produced 125 kg fruits (Godara *et al.,* 1980). Singh *et al.* (1983) reported the highest yield in *cv.,* Umran (142 kg/tree) and the lowest in *cv.,* Chhuhara (65 kg/tree).

Storage

Ber fruits are usually stored at ambient/room temperature (25–35 ^{0}C) from harvest until their consumption. Jawanda *et al.* (1980a) observed that Umran and Sanaur-2 fruits could be stored for up to 10–12 days at room temperature. Freshly harvested fruits can be stored in containers such as gunny, net or polythene bags, cloth packs and boxes for 4-15 days at room temperature (25-35 ^{0}C) without loss of organoleptic quality.

- ☆ Fruits can be stored in a cool chamber for 6-10 days. However, the high humidity, which develops in the cool chamber, is conducive to spoilage.
- ☆ Fruits can retain acceptable organoleptic quality for 3 weeks kept in polythene bags and baskets at 13 °C in an incubator.
- ☆ Kept in polythene bags in a cold storage at 10 ^{0}C, fruits of some cultivars can be stored for 28-42 days.
- ☆ Fruits can be stored frozen at -18 ^{0}C for 6 weeks.

The shelf-life depends mainly on the Ber cultivar, storage temperature, packaging method and stage of harvest. During storage, the fruits loose weight and shrivel, change colour and become red, loose acidity and ascorbic acid, but gain in sweetness.

The storage life of Ber fruits is extremely short and the rapid perishability of the fruits is a problem. At ambient temperature a shelf life of 2–4 days is common. Due to the surplus of fruits in the local markets during peak season, a substantial quantity goes to waste, resulting in heavy postharvest losses. A cost and returns analysis performed by (Gupta *et al.*, 1981) showed that Ber production is highly remunerative but requires proper handling with respect to pre-harvest, harvesting and postharvest treatments, packaging, transportation, storage, postharvest pathology, processing, *etc.* Profits could be enhanced if efforts to increase production are supplemented with efforts to minimize postharvest losses and enhance shelf life.

Ber fruits are usually stored at ambient/room temperature (25–35 °C) from harvest until their consumption. Panwar, 1981 reported that Ber fruits remained in marketable condition for about one week. Ripe fruits of Ber when stored at room temperature without any treatment remained for up to 7 days (Bal, 1982). Gupta *et al.*, observed that the shelf life at room temperature was the longest in Sanaur-5, followed by Ponda, Reshmi and Umran cultivars. Pareek and Gupta (1988) observed the shelf life of Gola and Kaithli cultivars at ambient temperature for up to 7 and 10 days, respectively. *Z. spinachristi* fruits could be stored for 6 days at room temperature (Abbas *et al.*, 1990; Al-Niami and Abbas, 1988). Golden yellow colour ripe fruits of Umran could be stored for about one week at 30 °C (Anonymous, 1990). The fruits of cultivar Gola were suitable for eating for up to 8 days of storage (Anonymous, 1990). Siddiqui and Gupta (1995) observed 38 per cent PLW, 35.3 per cent over ripening and 13.4 per cent decay loss in cultivar Umran at room temperature after 6 days of storage. In contrast to this Gupta and Kadam, 1995 reported that Ber fruits stored at ambient temperature had a short life of 3 days only. Under ambient conditions, Ber fruits showed a high degree of pathological infection and loss in colour, and could be stored for only 9 days (Vishal Nath *et al.*, 1999). Pareek *et al.* (1983) observed that the storage environment did not affect the levels of total sugars in the fruits. It was observed that Virosil Agro (2.5 and 5 per cent) and Bavistin (1 per cent) maintained the original levels of reducing sugars during the storage period.

Zero Energy Cool Chamber (ZECC)

High temperature and moderate humidity at the time of fruit maturity (February to March) leads to the attack of different micro-flora that caused decay, increased PLW and reduced shelf life and quality. These factors lead to heavy losses, which can be minimized by storing fruits in a zero energy cool chamber. ZECC have been designed to enhance the shelf life of fruits and vegetables by lowering the temperature and increasing the relative humidity inside the chambers *via* passive evaporative cooling (Roy and Khurdiya, 1986). Ber fruits of Kaithli, Umran and Gola cultivars could be stored in these chambers for 14, 15 and 18 days, respectively (Anonymous, 1988-1989). Siddiqui and Gupta (1989) stored fruits in these chambers up to 6–10 days. The fruits of cultivar Gola were found to be in acceptable condition up to 12 days of storage in ZECC. There was a decrease in fruit firmness, specific gravity and organoleptic score with a corresponding increase in acidity of fruits under ZECC (Ramkrishan and Godara, 1993). Mean PLW (10.36 per cent) was recorded for Umran cultivar stored in ZECC after 12 days of storage, which were organoleptically (Fageria, 1999; Dhaka, 2000). The significant reduction in PLW under cool chamber was due to prevailing higher humidity and lower temperature, which lowered the transpiration rate as well as ethylene production at the lower temperatures.

Cold Storage

Several studies have examined the effect of low temperature on postharvest changes in the chemical constituents of Ber fruits and on their storage behaviour (Jawanda *et al.,* 1980; Al-Niami and Abbas, 1988; Al-Niami *et al.,* 1989). Jawanda *et al.* (1980) observed that Umran and Sanaur-2 fruits could be stored for up to 30 and 40 days, respectively, in commercial cold storage (0–3.3 °C). Jain *et al.* (1981) reported that fruits stored in perforated polythene bags and baskets at 13 °C in biochemical oxygen demand (BOD) incubators remained at acceptable organoleptic quality for up to 3 weeks. Panwar (1981) stored *cv.,* Umran and Kaithli Ber up to 42 days at 10 °C. The shelf life of Gola and Kaithli cultivars of Ber at 1.7 °C was found to be 42 and 28 days, respectively (Pareek and Gupta, 1988). In cold storage (10 °C, 79 per cent relative humidity), fruits of cultivars Gola, Kaithii and Umran remained acceptable for up to 42, 28 and 35 days, respectively (Anonymous, 1989). The golden yellow coloured ripe fruits of Umran could be stored for about 3 weeks at low temperatures ranging from 0–4 °C (Anonymous, 1990). According to Monthira (1982) fruits could be stored in perforated polythene bags for 8, 16 and 24 days at 15 °C, 10 °C and 5 °C, respectively. Fruits stored at 5 °C lost only 48 per cent of their weight during the entire 12 weeks storage duration, while fruits stored at 22 and 15 °C lost 70 and 75 per cent of their weight, respectively. At 3 weeks of storage more than 40 per cent of fruits had shriveled under the 22 °C and 15 °C storage temperatures compared with only 3 per cent under the 5 °C storage temperature (Tembo *et al.,* 2008). The difference in storage life seems due to variation in year of production, regions, orchard management practices, irrigation geometry, maturity stages, locations of the fruits and the time of harvesting fruits from tree, and the storage environment, *etc.* Both Chinese jujube and *Z. spinachristi* can be freeze-dried or frozen although vitamin C content tends to decrease (Dzheneeva and Chernogorod, 1989; Al-Hijiya *et al.,* 1989). Fumigation to remove insects is possible for jujubes; use of hydrogen phosphide for this purpose has been demonstrated for fruits of Chinese jujube intended for herbal medicines (Khalid *et al.,* 1988).

Postharvest Treatments

To extend the shelf life and to reduce decay losses in storage Ber fruits can be treated as follows:

They can be dipped in cold water for 2 hours or exposed to cold air for 4 hours immediately after harvest to remove field heat. They can be dipped in calcium chloride or ascorbic acid solution. They can be treated with growth regulators (*e.g.* Cycocel), waxed or fumigated. They can be sprayed with fungicides, *e.g.* thiobendazole (500 ppm) or 0.2 per cent $ZnSO_4$. Fungicides should only be used to provide certain health and safety regulations.

Composition of Fresh Ber Fruit

Food Value/100 g Edible Portions

Constituent	*Range*
Moisture (per cent)	79-82
Starch (per cent)	0.72-1.15
TSS (^{o}Brix)	12-21
Total sugars (per cent)	3.1-4.5
Reducing sugars (per cent)	1.4-9.7
Non-reducing sugars (per cent)	1.3-9.7
pH	4.2-4.9
Acidity (per cent)	0.13-1.42
Protein (per cent)	0.96-1.75
Total Ash (per cent)	0.34-0.45
Ascorbic acid (mg/100gm)	39-166

Uses

1. The fruit is eaten raw or pickled or used in beverages. It is quite nutritious and rich in Vitamin C.
2. It is second only to guava and much higher than citrus or apples. In India, the ripe fruits are mostly consumed raw, but are sometimes stewed. Slightly under ripe fruits are candied by a process of pricking, immersing in a salt solution.
3. Ripe fruits are preserved by sun drying and a powder is prepared for out-of-season purposes. It contains 20 to 30 per cent sugar, up to 2.5 per cent protein and 12.8 per cent carbohydrates.
4. Fruits are also eaten in other forms, such as dried, candied, pickled, as juice, or as Ber butter. The dried ripe fruit is a mild laxative.
5. The seeds are sedative and are taken, sometimes with buttermilk, to halt nausea, vomiting, and abdominal pains in pregnancy. They check diarrhea, and are poultice on wounds.
6. The fruit is believed to purify blood and to help in digestion. The bark is said to be a remedy in diarrhoea. The root is used as decoction in fever and as powder applied to ulcers and old wounds. The leaves form a plaster in strangury and are used in conjunctivitis.
7. According to Ayurved, the bitter *Z. jujuba* root and leaves are cooling and cure kapha, biliousness and headache. The bark cures boils and is good in dysentery and diarrhoea. The unripe fruit removes vata but causes Cough.

8. According to Yunani medicine, the root and bark are tonic, the leaves are anthelmintic and good in stomatitis, gum bleeding, wound, syphilitic ulcers, asthma and liver complaints; the flowers afford a good collyrium in eye trouble.
9. The fruits of *Z. nummularia* are considered appetizer and stomachic in Ayurveda. In Yunani system its leaves are used in scabies and boils and for smoking to cure colds. The leaf decoction is used as hip bath for joint pains and as gargle in sore throat and in bleeding gums.

Value Added Products

Extensive studies have been carried out using Ber fruits to prepare various processed products, such as candy (Pareek, 1983; Gupta *et al.,* 1980; Gupta and Kadam, 1995), dehydrated products (Gupta and Kadam, 1995, Agrawal *et al.,* 1997), juice and wine (Gupta and Kadam, 1995; Khurdia, 1980), jam and jelly (Gupta *et al.,* 1981), and shreds and powder (Patil *et al.,* 1999). Ber fruits are highly nutritious, rich in ascorbic acid and contain fairly good amount of vitamin A and B, minerals like calcium, phosphorus and iron. Ber fruits are also higher in calorific value and ascorbic acid content than the orange (Jawanda and Bal, 1978). Most fruits are seasonal and highly perishable and it is estimated that the total loss of fruits in India, for want of adequate postharvest care, transportation and storage is around 20-30 per cent (Madan and Ullasa, 1993).

It is a fruit of Indian origin, which finds a place in the ancient Indian scriptures (Parek and Vashistha, 1983). It is a rich source of ascorbic acid. Further, cost of its production is low and therefore, it is considered poor man's apple (Kudachikar *et al.,* 2000). Ber is consumed mainly as a fresh fruit and small quantities are processed. It can be processed into delicious products such as juice (Wasker and Garande, 1999), RTS beverage (Khurdia 1980), chutney and pickles (Bal and Singh 1978), jam, jellies, dried fruit and powdered pulp (Sood *et al.,* 1980) and candy (Unde *et al.,* 1998). Dried Bers can be used in the preparation of various Indian delicious like halwas, pinnies, laddoos, ice cream, pudding, fruit chats and fruit salads (Aggarwal *et al.,* 1997). Ber fruits may be processed into several products, such as jam, jelly, pulp, juice, powder, dried Ber, candy, slices, tutti-frutti and wine. The by-products of the processing industries may be utilized for the extraction of pectin.

Juice and Pulp

Juicy varieties of the Ber are better suited for preparing the juice (Khurdya and Singh 1975). A process for the extraction of juice from the Ber fruits has been standardized (Kadam *et al.,* 1991). The juice is extracted, allowed to stand at 4 ^{0}C for two hours. After removal of the scum, juice is passed through four layers of the muslin cloth. Then, it is filtered through an ordinary filter paper. The recovery of the juice is about 40 per cent of the fresh weight. The Ber pulp can serve as a base material for the preparation of nectar, leather *etc.* (Khurdiya and Singh, 1975).

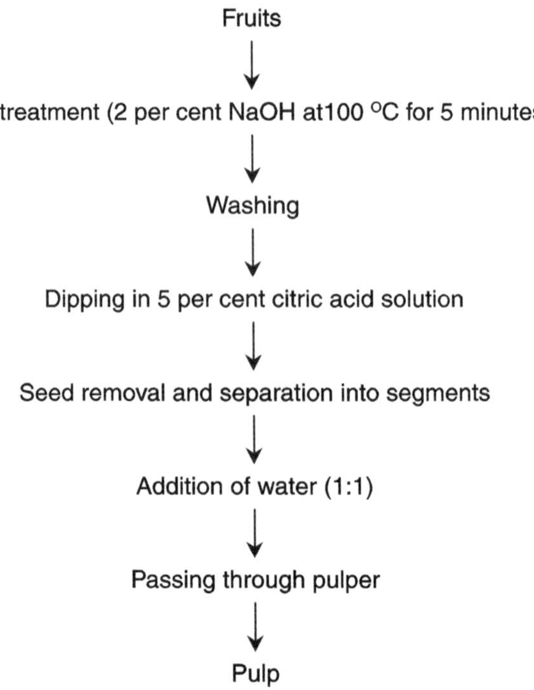

Ber Chutney

Dried Ber, fresh fruits or pulp can be used to prepare a spicy fruit chutney. After cleaning and sorting, the fruit is soaked overnight in water to rehydrate it. Usually this length of time is sufficient to soften the fruit but if any remains hard after soaking, the fruit should be heated. The soaked fruit is removed from the soaking water, rinsed and placed in a pan with sugar at 0.7 to 1 kg per kg dried fruit, according to taste. The Ber and sugar are stirred to dissolve the fruit and then heated for 10 to 15 minutes until the total soluble solids content is 55 °Brix, measured with a Refractometer. Spices such as chilli powder, cumin, aniseed, ground cinnamon, ground nutmeg and black pepper are dried, roasted in a heavy iron pan before use. Then they added with mustard powder to the heated Ber fruits, stirred well to mix and heated. Acetic acid (vinegar) and salt are added, stirred well to mix and heated again. Heating is continued until the TSS reaches 60 °Brix. The chutney

Sl.No.	*Ingredients*	*Quantity*	*Sl.No.*	*Ingredients*	*Quantity*
1	Ber (dried)	1 kg	7	Cinnamon	2 g
2	Sugar	1 kg	8	Cardamom	1 g
3	Chilli powder	5 g	9	Nutmeg	1 g
4	Mustard powder	10 g	10	Black pepper	1 g
5	Cumin	2 g	11	Acetic acid	5 ml
6	Aniseed	2 g	12	Salt	30 g

is then ready for bottling. A preservative, potassium or sodium metabisulphite (KMS or NaMS) can be added as an option at this stage if desired.

Ready to Serve

A good quality RTS may be prepared from Ber juice at relatively low cost Ber juice (10 per cent) can be used for ready to serve beverage (Kadam *et al.,* 1991). Carbonated beverage of Ber was highly acceptable and has excellent keeping quality for three months (Khurdiya 1989 and 1990).

Flow-Chart for Processing of RTS Beverages

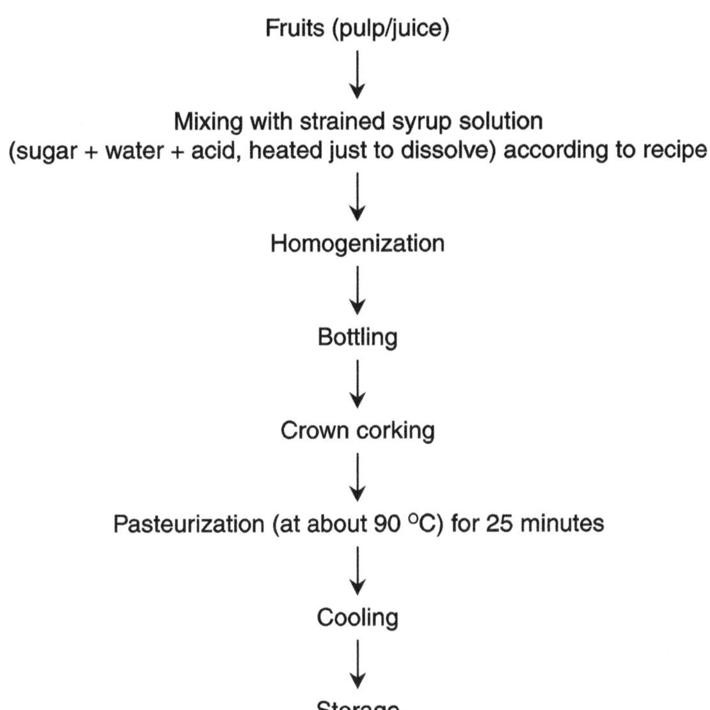

Dried Ber and Powder

Excellent quality Ber was prepared by treating fruits with sulphur dioxide at 3.5-10 gm/kg for 3 hours fallowed by drying in sun for 7-10 days or in a cabinet drier at 60 °C for 20-35 hours until the moisture content reduces to 15 per cent (Khurdya and Singh, 1975). The product may be consumed as such or it may be reconstituted in10 per cent sugar solution to be consumed as a liquid beverage. A process for Ber fruits drying has been developed (Chavan *et al.,* 1992) from which powder has successfully been prepared. In western India, powder of *Z. nummularia* is used for making churan.

Technical Flow-Chart for Processing of Dehydrated Fruit (Ber)

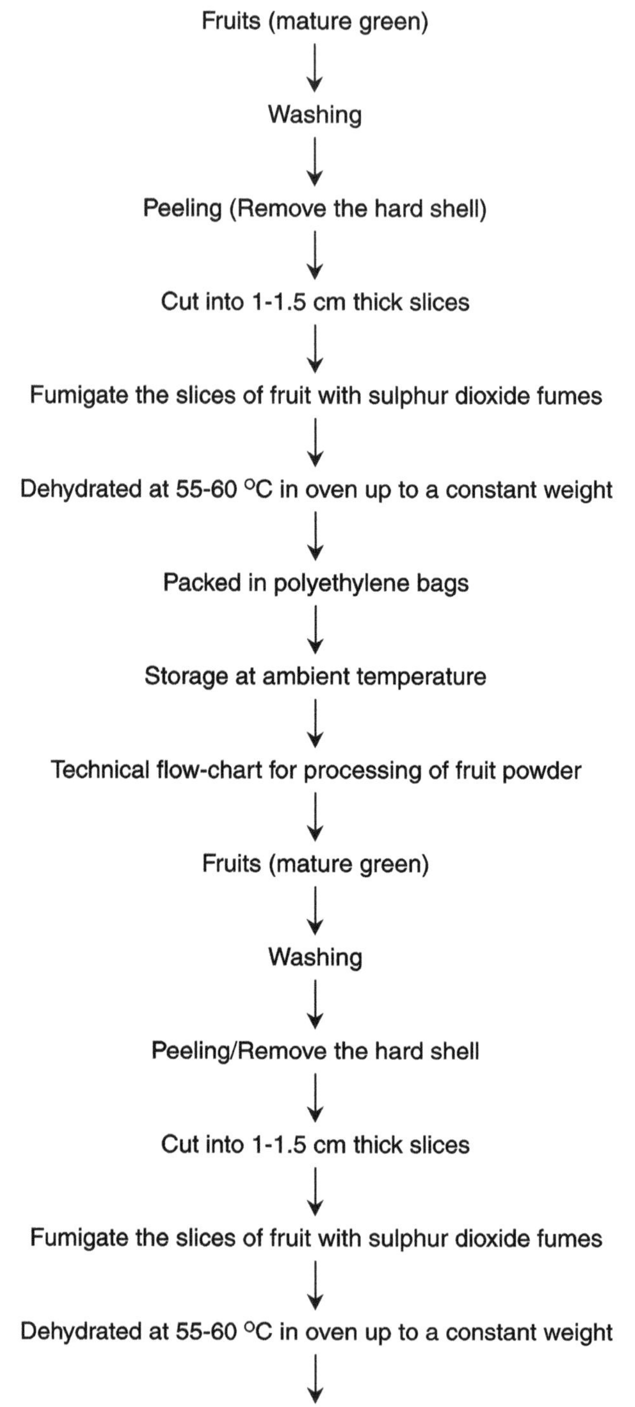

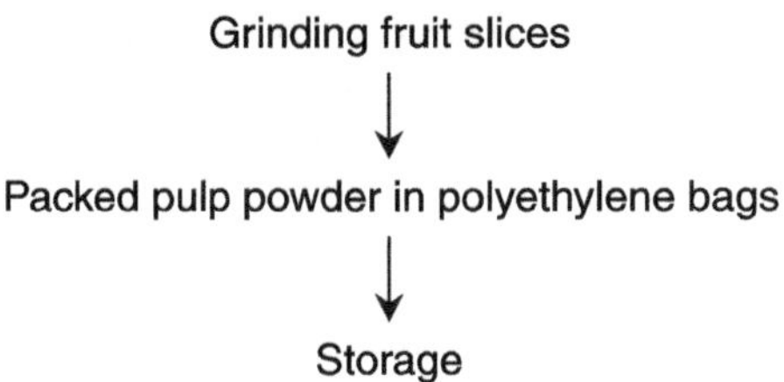

Ber Preserve

Mature fruits can be used to make a preserve, known locally in India as murabba. Fruits of the cultivars Umran, Banarsi, Karaka and Kaithli are the best for the preparation of preserve (Pareek, 2001). The best preserve is made from fully mature fruits that are at the hard stage. Ripe fruits are not suitable since the structure will be too soft. After washing and sorting the fruits, the skins need to be pricked so that sugar can impregnate the fruit. They are either pricked with a fork or by using a pricking board or pricking machine, similar to the one that is used for pricking olives. The pricked fruits then need to be softened. One method of softening the fruits is to soak them in brine solutions of gradually increasing concentration. For example, 2 per cent brine for the first day, 4 per cent brine the second day, 6 per cent brine on the third day and then 8 per cent brine solution for one to three months. However, this method proved to be too time consuming and it was difficult to remove all traces of salt from the fruit (Pareek, 2001). Therefore alternative methods of softening have been used. A simpler method of softening the fruit is to blanch the pricked fruits by plunging them into boiling water for anything from 2 to 190 minutes depending upon the cultivar, stage of maturity and size of fruit. After blanching, the fruits are dipped in cold water to stop the process. At this stage, the fruit can be peeled if desired, by using a pectinase enzyme (fruits are dipped in a solution of 2.5 per cent pectinase enzyme for 72 hours). This is an optional stage.

There are many different methods of preserving Ber, and they differ in the length of time it takes to reach the end point at 70 per cent total solids. The following method is one that is recommended by Khurdiya and Singh (1975):

- ☆ Day 1. Immerse the fruits in a 30 ^{0}Brix sugar solution. Add 0.5 per cent citric acid and leave overnight.
- ☆ Day 2. Remove the fruits from the syrup, add sugar (about 250 g/kg fruit) to the syrup and boil to dissolve the sugar. Allow the syrup to cool and replace the fruits into it. Leave for two days.
- ☆ Day 4. Remove the fruits from the syrup, add sugar (about 250 g/kg fruit) to the syrup and boil to dissolve the sugar. Allow the syrup to cool and replace the fruits into it. Leave for two days.

- ☆ Day 6. Remove the fruits from the syrup, add sugar (about 250 g/kg fruit) to the syrup and boil to dissolve the sugar. Allow the syrup to cool and replace the fruits into it. Leave for two days.
- ☆ Day 8. Measure the TSS of the product. It should be between 65 and 70 ^{0}Brix. Bottle the preserved fruit in the 70 ^{0}Brix syrup. The total soluble solids (TSS) content of the fruits becomes equalized at 65 to 70 ^{0}Brix.

Candy and Tutti-Fruity

Earlier attempts to prepare the candy resulted in having dark brown coloured product (Gharate, 1984; Gupta *et al.,* 1980; Gupta *et al.,* 1981). Recently a method of preparing Ber candy has been standardized to obtain a product with an attractive golden yellow colour and better storage life (Kadam *et al.,* 1991). Kumar *et al.* (1992) evaluated four candy making recipes and found that the use of 75 per cent TSS resulted in highest organoleptic score. Candy of cv. Umran prepared using this recipe was stored for 10 months at room temperature and its stability determined. The Total Soluble Solids, acidity and browning increased whilst the ascorbic acid content and organoleptic value decrease during storage. Fruits of the cultivar Illaichi have the best organoleptic properties for candy making. Kaithli, Kathaphal, Umran and Narma are also good cultivars for candy making. Fruits should have a high total soluble solids (TSS) content, low acidity and be of average size (Pareek, 2001).

Wine

Ber juice can also be used for the preparation of wine. Kainsa and Gupta (1979) reported the preparation of wine from the Ber pulp. Adsule *et al.* (1993) also reported that good quality wine prepared from Ber juice. There is a mention of use of an intoxicating drink made from jujube fruits in *Vishnu Dharma Sutra* (22-83-84).

Technical Flow Sheet for Processing of Fruit Wine

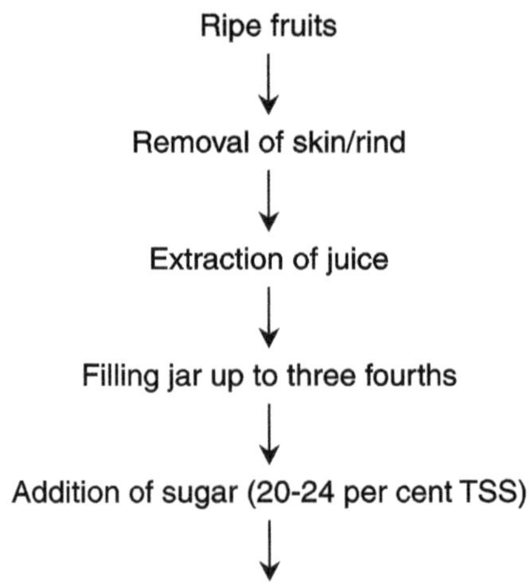

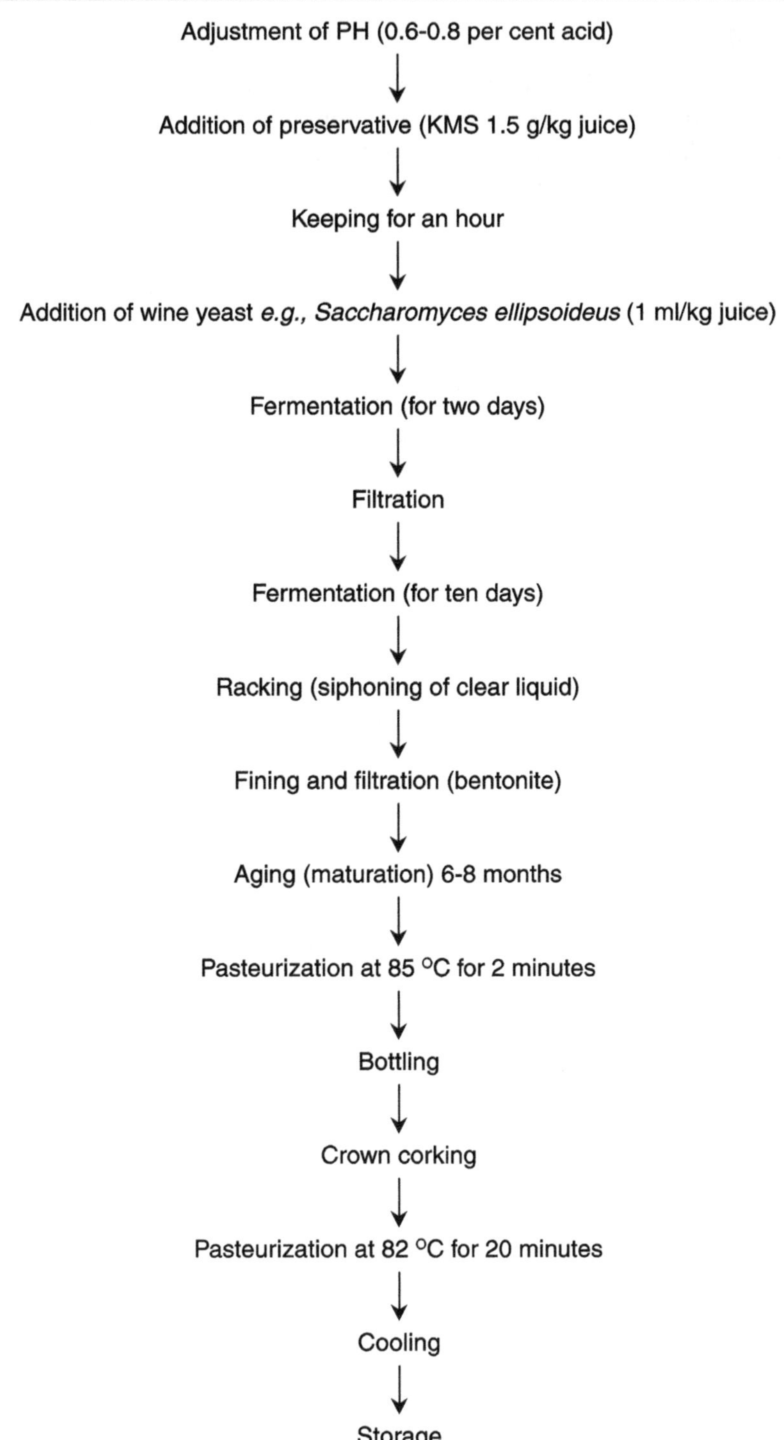
Adjustment of PH (0.6-0.8 per cent acid)
Addition of preservative (KMS 1.5 g/kg juice)
Keeping for an hour
Addition of wine yeast e.g., Saccharomyces ellipsoideus (1 ml/kg juice)
Fermentation (for two days)
Filtration
Fermentation (for ten days)
Racking (siphoning of clear liquid)
Fining and filtration (bentonite)
Aging (maturation) 6-8 months
Pasteurization at 85 °C for 2 minutes
Bottling
Crown corking
Pasteurization at 82 °C for 20 minutes
Cooling
Storage

Jam and Jelly

Ripe Ber fruits can be used to make jam. Preserved Ber pulp can also be used as the starting material for Ber jam. An excellent jelly can be prepared from unripe Ber fruits by adding proper amount of pectin and acid to this fruit or by mixing it with other fruits (Morton and Morton, 1955). The jam prepared from Ber fruits was leathery due to the presence of mucilaginous compound in the fruits. Jam with

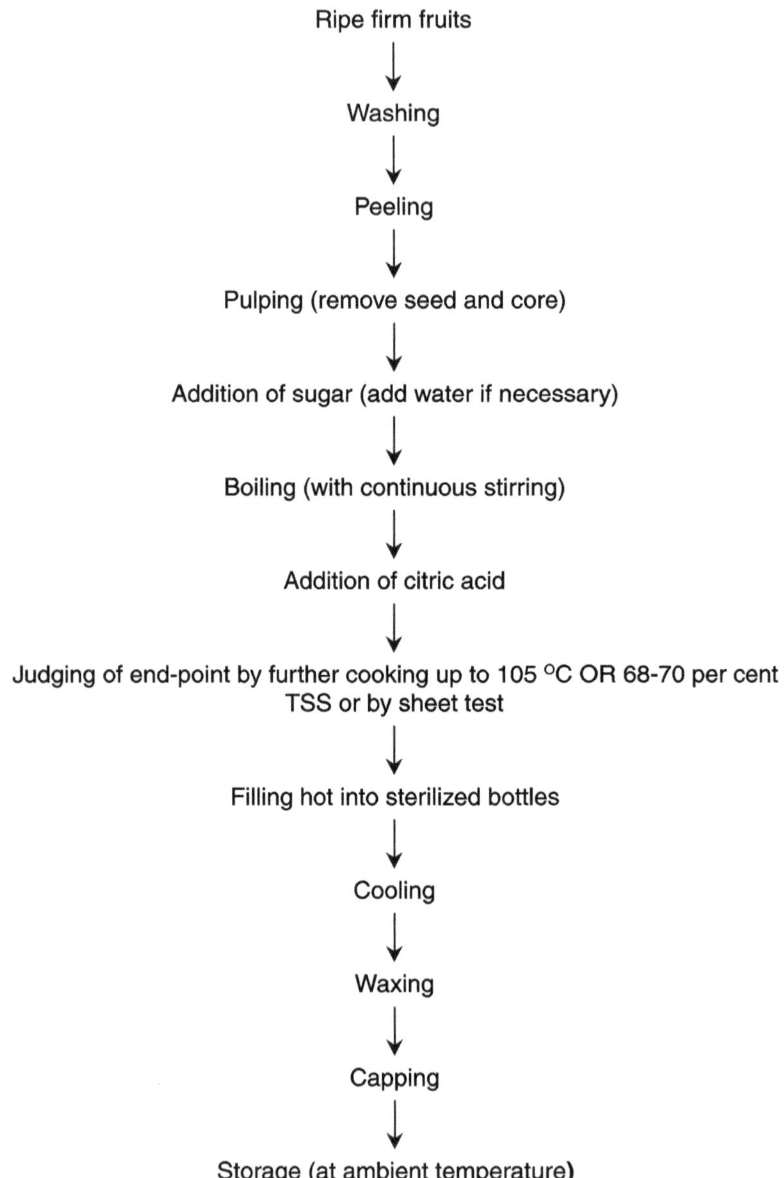

the best texture and taste can prepared from Umran fruits, without the peel on and with 1 per cent added pectin. More specifically, they found that the best recipe for Ber jam included 750 g sugar per kg pulp and with an acidity of 0.75 per cent (Dawney and Kulwal, 2002).

Technical Flow-Chart for Processing of Jelly

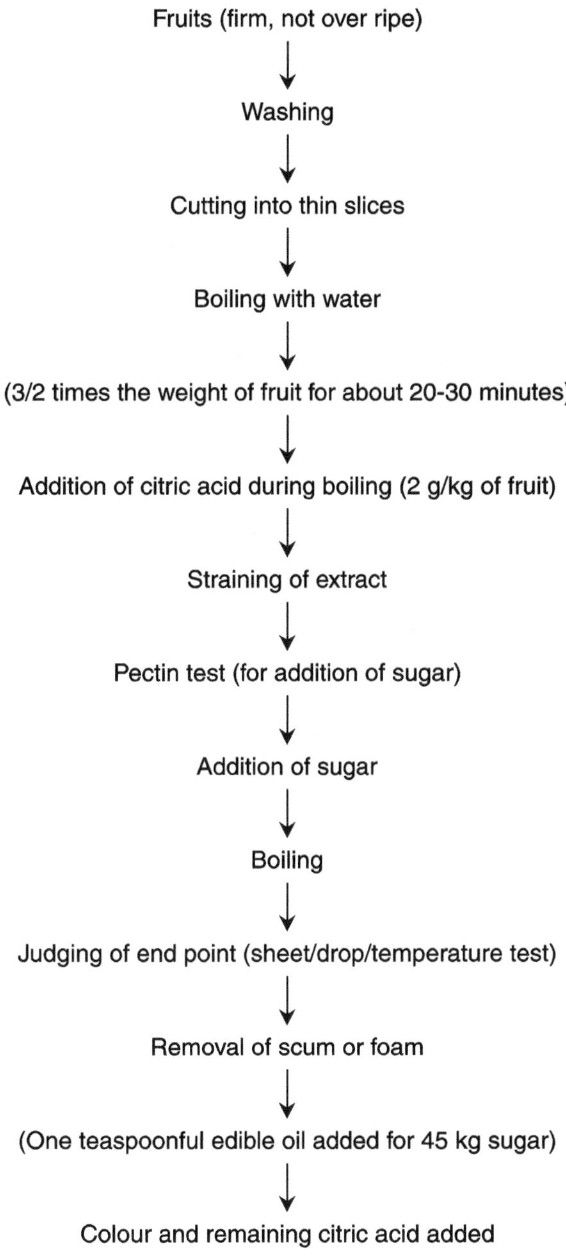

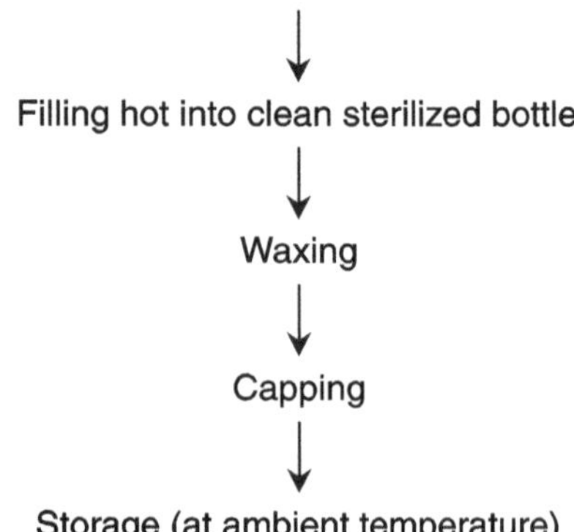

Candied Jujubes and Jujube Syrup (California Rare Fruit Growers, 1996)

Wash about 1.5 kg-dried jujubes, drain and prick each several times with a fork. In a kettle bring to boil 5 cups water, $5^{1}/_{2}$ cups sugar, and 1-tablespoon corn starch. Add the jujubes and simmer, uncovered, stirring occasionally for 30 minutes. Cool, cover and chill overnight. Next day bring mixture to a boil and simmer, uncovered, for 30 minutes. With a slotted spoon lift jujubes from syrup and place slightly apart on rimmed pans. The fruit (jujubes) should be dried in an oven or in sun until date-like, turning occasionally. Boil the syrup remaining from the candied jujubes, uncovered, until reduced to about two cups.

Jujube Butter (USDA 1980)

3.5 ltr jujube pulp, 1-teaspoon nutmeg, 1/2-teaspoon cloves, 2 teaspoons cinnamon, 2.75 ltr sugar, 125 ml vinegar, 1 lemon *etc.*

Boil fruit until tender in sufficient water to cover it. Rub cooked fruit through a sieve or colander to remove the skin and seeds. Cook slowly until thick, put in jars and seal while hot.

REFERENCES

Abbas, M. F., Al-Niami, J. H. and Asker, M. A. (1990). The effect of temperature on certain chemical constituents and storage behaviour of jujube fruits (*Ziziphus spina christi* L. Willd.) *cv.* Zaytani. *Haryana Journal of Horticultural Science.* 19: 263–267.

Adsule, R. N., Chougule, B. A., Kotecha, P. M. and Kadam, S. S. (1993). Processing of Ber II. Preparation of wine. *Beverage and Food World.* 19: 16.

Agarawal, P., Kaur, B. and Bal, J. S. (1997). Studies on dehydration of different cultivars for making Ber chuharas. *Journal of Food Science and Technology.* 34: 534-536.

Al-Hijiya, M. N., Ali, S. H. and Al-Delaimy, K. S. (1989) Proximate composition and the effect of freezing on ascorbic acid content of *Ziziphus spina-christi* fruit. *Iraqi Journal of Science.* 30(1): 73-78.

Al-Niami, J. H. and Abbas, M. F. (1988). The effect of temperature on certain chemical changes and the storage behaviour of jujube fruits (*Ziziphus spina- christi* L. Willd.). *Journal of Horticultural Science.* 63:723–724.

Al-Niami, J. H., Abbas, M. F. and Asker, M. A. (1989). The effect of temperature on some chemical constituents and storage behaviour of jujube fruit *cv.* Zayoui. *Basrah Journal of Agricultural Science.* 2:31–36.

Anonymous (1988–89). Annual Progress Report. All India Coordinated Research Project on Postharvest Technology of Horticultural Crops. *Indian Council of Agricultural Research,* New Delhi, India. 1989.

Anonymous (1990). Annual Report, Central Food Technological Research Institute, Mysore, India.

Bailey, L. H. (1947). The standard cyclopaedia of Horticulture. Macmillan and Company, New York.

Bakhshi, J. C. and Singh, P. (1974). The Ber- a good choice for semi-arid and marginal soils. *Indian Horticulture.* 19: 27-30.

Bal, J. S. (1982). A study on biochemical changes during room and refrigerator storage of Ber. *Progressive Horticulture.* 14: 158–161.

Bal, J. S. and Chohan, G. S. (1981). Fixation of maturity standards of Ber cv. Umran. *Punjab Horticulture Journal.* 21(1): 70-75.

Bal, J. S. and Singh, P. (1978). Developmental physiology of Ber (*Ziziphus mauritiana* Lamk.) var. Umran. I. Physical changes. *Indian Food Packer.* 32:59–61.

Bal, J. S., Bajwa, G. S. and Singh, P. (1995). Effect of ethephon application at turning stage on ripening and quality of Umran Ber. *Punjab Horticultural Journal.* 33(1/4): 84-87.

Bal, J. S., Jawanda, J. S. and Kahlon, P. S. (1995a). Effect of preharvest spray of growth regulators on the pectin methylesterase activity of Ber fruit during cold storage. *Journal of Food Technology.* 33: 159–161.

Bhatia, S. K. and O. P. Gupta. (1985). Chemical changes during development and ripening of Ber fruits. *Punjab Horticultural Journal.* 25: 62-66

Brandis, D. (1906). Indian trees. An account of trees, shrubs, woody climBers, bamboos and palms indigenous or commonly cultivated in British India. Bishen Singh Mahendra Pal Singh, Dehradun, India. : 169–172.

California Rare Fruit Growers (1996) http://www.crfg.org/pubs/ff/jujube.html

Chavan, U. D., Adsule, R. N. and Kadam, S. S. (1992). Processing of Ber III. Preparation of Ber powder and tutti-frutti. *Beverage and Food World.* 20: 28-29.

Dawney, T. and Kulwal, L. V. (2002). Preparation of jam from Ber fruits (*Ziziphus mauritiana* L.). *Indian Food Packer.* January-February, 57-59.

Dhaka, R. S., Lal, G., Fageria, M. S. and Agrawal, M. (2000). Studies on zero energy cool chamBer for storage of Ber (*Ziziphus mauritiana* Lamk.) fruits under semi-arid conditions. *Annals of Arid Zone.* 39:439–441.

Dzheneeva, E. L. and Chernogorod, L. B. (1989) Deep freezing of *Ziziphus* [in Russian]. *Sbornik Nauchnykh Trudov - Gosudarstvennyl Nikitskil Botanicheskil Sada,* No. 109: 119-127.

Fageria, M. S., Dhaka, R. S., Sharma, B. M. and Gujar, K. D. (1999). Effect of time of harvest on postharvest behaviour of Ber fruits *cv.* Mundia under semi-arid conditions. *In*: Faroda, A. S., Joshi, N. L., Kathiya, S. and Kar, A. (Eds.) Management of Arid Ecosystem. Arid Zone Research Association of India and Scientific Publisher, Jodhpur, India. 365–368.

Gharte, A. N. (1984). Studies on preparation of candy from Ber. M.Sc. (Ag.) Thesis submitted to MPAU, Rahuri, India.

Godara, N. R., Chauhan, K. S. and Bisla, S. S. (1980). Evaluation of mid-season ripening Ber (*Ziziphus mauritiana* Lamk.) germplasm. *Haryana Journal of Horticultural Science.* 9(3/4): 101-105.

Gupta, O. P. and Kadam, S. S. (1995). Ber. *In*: Salunkhe, D. K. and Kadam, S. S. (Eds) Handbook of Fruit Science and Technology. Marcel Dekker Inc New York.: 387.

Gupta, O. P., Kainsa, R. L. and Chauhan, K. S. (1980). Postharvest studies on Ber fruits (*Ziziphus mauritiana* Lamk.). 1. Preparation of candy. *Haryana Agricultural University Journal of Research.* 10:163.

Gupta, O. P., Kainsa, R. L., Chauhan, K. S. and Dhawan, S. S. (1981). Postharvest studies in Ber fruits (*Ziziphus mauritiana* Lamk.) IV. Comparison of sugar and gur for the preparation of candies. *Haryana Agricultural University Journal of Research.* 1: 369–392.

Gupta, O. P., Siddiqui, S. and Chauhan, K. S. (1987). Evaluation of various calcium compounds for increasing the shelf life of Ber (*Ziziphus mauritiana* Lamk.) fruit. *Indian Journal of Agricultural Research.* 21: 65–70.

Gupta, O. P., Siddiqui, S. and Gupta, A. K. (1989). Effect of preharvest sprays of various chemicals on the storage of Ber (*Ziziphus mauritiana* Lamk.). *Research and Development Reporter.* 6: 35–40.

Jain, R. K. and Chitkara, S. D. (1978). Effect of packings in transportation on quality of Ber fruit (*Zizyphus mauritiana* Lamk.). Inter. Symp. Arid Zone Res. Development, 14-16th Feb., Jodhpur.

Jain, R. K., Chitkara, S. D. and Chauhan, K. S. (1981). Studies on physico-chemical characters of Ber (*Ziziphus mauritiana* Lamk.) fruits *cv.* Umran in cold storage. *Haryana Journal of Horticultural Science.* 106: 141–146.

Jawanda, J. S. and Bal, J. S. (1978). The Ber highly paying and rich in food value. *Indian Horticulture*. 23: 19-21.

Jawanda, J. S., Bal, J. S., Josan, J. S. and Mann, S. S. (1980). Studies on the storage of Ber fruits II. Cool temperature. *Punjab Horticultural Journal*. 20: 171- 178.

Jawanda, J. S., Bal, J. S., Josan, J.S. and Mann, S. S. (1980a). Studies on the storage of Ber fruits. I Room temperature. *Punjab Horticultural Journal*. 20: 56–61.

Johnston, M. C. (1972). Rhamnaceae. In: Milne RE and Polhill RM (Ed), Flora of Tropical East Africa. Crown Agents, London.

Kadam, S. S., Chavan, U. D. and Dhotre, V. A. (1991). Processing of Ber 1. Preparation of ready-to-serve beverage and candy. *Beverage and Food World*. 18: 13-15.

Kainsa, R. L. and Gupta, O. P. (1979). Postharvest studies on Ber fruits (*Zyziphus mauritiana* Lamk.) II. Preparation of wine. *Haryana Agricultural University Research Journal*. 9: 260.

Khalid, Q., Sultana, L., Basit, N., Mahmood, I., Siddiqui, M. A. and Beg, M. A. A. (1988). Studies on the preservation of herbal medicines by treatment with hydrogen phosphide. *Pakistan Journal of Scientific and Industrial Research*. 31(12): 837-841.

Khera, A. P. and Singh, J. P. (1976). Chemical composition of some Ber cultivars (*Ziziphus mauritiana* L.). *Haryana Journal of Horticultural Science*. 5(1): 21–24.

Khurdia, D. S. (1980). Studies on dehydration of Ber (*Zyziphus mauritiana* Lam) fruit. *Journal of Food Science and Technology*. 17: 127-130.

Khurdiya, D. S. (1990). A study on fruit juice based carbonated drink. *Indian Food Pack*. 44: 45.

Khurdiya, D. S. and Singh, R. M. (1975). Ber and its products. *Indian Horticulture*. 20: 5-25.

Khurdiya, D. S. (1989). Carbonation in fruit beverages 1. *Beverage and Food World*. 16: 9.

Kudachikar, V. B., Ramana, K. V. R. and Eipeson, W. E. (2000). Pre and postharvest factors influencing the shelf life on Ber (*Ziziphus mauritiana* Lamk.): A review. *Indian Food Packer*. Jan.–Feb.: 81–90.

Kumar, S., Ojha, C. M., Bhagwan, D., Awasthi, O. P. and Nainwal, N. C. (1992) Potentiality of Ber (*Ziziphus mauritiana* Lamk.) cultivars for candy making. *Progressive Horticulture*. 24(1-2): 74-78.

Lin, M. J. and Cheng, C. Y. (1995). A taxonomic study of the genus *Ziziphus*. *Acta Horticulture*. 390:161–165.

Lyrene, P. M. (1979). The jujube tree (*Ziziphus jujuba* Lamk.). *Fruit Varieties Journal*. 33(3): 100-104.

Madan, M. and Ullasa, B. A. (1993). Postharvest losses in fruits. *Advances in Horticulture* (eds. Chadha, K. L. and Pareek, O. P.), Malhotra Publishing House, New Delhi, 4:1795-1810.

Monthira, S. (1982). Effect of low temperature and packaging material on storage life of Bombay jujube (*Ziziphus mauritiana* Lamk.). Bangkok, Thailand.

Morton, K. and Morton, J. (1955). Fifty tropical fruits of Nassau. Text House, Coral Gables, New Delhi, India.

Panwar, J. K. (1981). Postharvest physiology and storage behaviour of Ber fruit (*Ziziphus mauritiana* Lamk.) in relation to temperature and various treatments. Thesis Abstracts, Haryana Agricultural University, Hisar. 7:64–65.

Pareek, O. P. (1983). The Ber. Indian Council of Agricultural Research. New Delhi, India.

Pareek, O. P. (2001). Ber. International Centre for Underutilized Crops, Southampton, UK :13.

Pareek, O. P. and Gupta, O. P. (1988). Packaging of Ber, datepalm and phalsa. In: A souvenir on packaging of fruits and vegetables in India. Agri-Horti Society, Hyderabad, India. 91–103.

Pareek, S., Fageria, M. S. and Dhaka, R. S. (2002). Performance of Ber genotypes under arid condition. *Current Agriculture*. 26: 63–65.

Parek, O. P. and Vashistha, B. B. (1983). Delicious Ber varieties of Rajasthan. *Indian Horticulture*. 28:13-16.

Patil, D. M., Katecha, P. M. and Kadam, S. S. (1999). Drying of Ber preparation of shreds and powder. Processed Food Industry. August: 14–15.

Raja, P. V. (1985). Flowering behaviour of Ber. M. Sc. Thesis, Indian Agricultural Research Institute, New Delhi, India.

Ramkrishan, N. R. and Godara, R. K. (1993). Physical and chemical parameters as affected by various storage conditions during storage of Gola Ber (*Ziziphus mauritiana* Lamk.) fruits. *Progressive Horticulture*. 25: 60–65.

Roy, S. K. and Khurdiya, D. S. (1986). Studies on evaporative zero energy input cooled chamBers for storage of horticultural produce. *Indian Food Packer*. 40: 26–31.

Sebastian, M. K and Bhandari, M. M. (1990). Edible wild plants of the forest areas of Rajasthan, India. *Journal of Economic and Taxonomic Botany*. 14: 689–694.

Siddiqui, S. and Gupta, O. P. (1989). Determination of maturity standards of Ber (*Ziziphus mauritiana* Lamk.) fruits. Haryana Agricultural University *Journal of Research*. 19(3): 260-262.

Siddiqui, S. and Gupta, O. P. (1989a). Effect of preharvest spray of calcium on shelf life of Ber (*Ziziphus mauritiana* Lamk.) fruits. *Research and Development Reporter*. 6: 172–176.

Siddiqui, S., Gupta, O. P. and Yamdagni, R. (1989). Effect of preharvest sprays of chemicals on the shelf life of Ber (*Ziziphus mauritiana* Lamk.) fruits *cv*. Umran. *Haryana Journal of Horticultural Science.* 18: 177–183.

Singh, R. R., Jain, R. K. and Chauhan, K. S. (1983) Flowering and fruiting behaviour of Ber (*Ziziphus mauritiana* Lamk.) under Gurgaon conditions. *Haryana Agricultural University Journal of Research.* 13(1): 112-114.

Sood, D. R., Wagle, D. S., Nainawatee, H. S. and Srivastav, H. C. (1980). Quality attributes of some Bers (*Ziziphus jujuba*) strains. *The Indian Journal of Nutrition and Dietetics.* 17: 447-451.

Tembo, L., Chiteka, Z. A., Kadzere, I., Akinnifesi, F. K. and Tagwira, F. (2008). Storage temperature affects fruit quality attributes of Ber (*Ziziphus mauritiana* Lamk.) in Zimbabwe. *African Journal of Biotechnology.* 7(8): 3092– 3099.

Unde, P. A., Kanwade, V. L. and Jadhav, S. B. (1998). Effect of syruping and drying methods of quality methods of Ber candy. *Journal of Food Science and Technology.* 35: 259-261.

Vishal Nath, Bhargava, R. and Pareek, O. P. (1999). Improvement in shelf life of Ber (*Ziziphus mauritiana* Lamk.) by postharvest treatments. 4th Agricultural Science Congress, Jaipur, India. February 21–24, 210.

Wasker, D. P. and Grande, V. K. (1999). Standardization of a method of juice extraction from Ber fruit. *Journal of Food Science and Technology.* 36: 540-541.

Yamadagni, R. (1985). Ber. In: *Fruits of India, Tropical and Subtropical* (ed. By Bose, T.K.), Naya Prakash, Calcutta, India.

4

Carambola

Scientific Name: *Averrhoa carambola*

Family: *Oxalidaceae*

Fresh fruits and vegetables are very important part of our lives. They are also an important source of health promoting agents and known as protective foods as they provide vitamins, minerals and fiber required for maintaining good health (Kanwar and Budhwar, 2018). The climatic condition of India is diversified way, therefore it provides the availability of fresh fruits and vegetables throughout the year. According to Press Information Bureau, Ministry of Agriculture, Government of India, India is the second largest producer of fruits in the world *i.e.* 82.632 MT after china with 154.364 MT. The postharvest losses during handling, transportation, storage, processing and distribution of fruits in India are between 17 to 35 per cent. *Averrhoa Carambola*, a member of the Oxiladacea family is a fruit distinguished for its unique and attractive star shape, therefore gaining a high demand in the international market, even though it is low risk specie. The fruit, with an undetermined origin, has traveled through various regions and gained multiple names according to its shape and sweetness such as, star fruit or sweet belimbing.

The word Carambola is derived from the Sanskrit word karmaranga meaning "food appetizer". It is a small and slow-growing tree cultivated in tropical and subtropical regions for its fruit and medicinal properties. The fruit is believed to have originated in South East Asia, Indonesia or Malaysia, but it is now cultivated throughout the tropics and subtropics of the world. The major Carambola producing countries are China, Taiwan, Malaysia, Thailand, Pakistan, Indonesia, Australia, West Indies and USA (Bose *et al.,* 2002). In India, it is distributed in Uttar Pradesh, Karnataka, Assam, West Bengal, Madhya Pradesh, Bihar and Tamil Nadu (Srivastava and Rajput, 2003). Carambola fruit features light green to yellow with attractive smooth waxy surface and weighs about 70-130 g. Inside, it's crispy, and juicy pulp can either be mildly sweet or extremely sour depending upon the cultivar type and amount of oxalic acid concentration. In some seed varieties, 2-5 tiny edible seeds found at the center of each angled cavity.

Use

It is a good source of potassium, copper, as well as folate and pantothenic acid. The Ascorbic acid level of the star fruit is believed to be responsible for its sweet or sour taste. For a sweet fruit, the ascorbic acid level is around 10.40 mg/100 ml of juice (Chang, 2002; Monolisa *et al.,* 2014). According to USDA (2005) Star fruit is rich in nutritive value, which contains 0.38 g Protein, 9.38 g Carbohydrates, 35.7 KJ Calories, 0.08 g Fat, 89-91 g Moisture, 0.80-0.90 g Fiber, 0.26-0.40 g Ash, 26-53.1 mg Ascorbic Acid, 4.4-6 mg Calcium, 0.32-1.65 mg Iron, 15.5-21.0 mg Phosphorus, 0.003-0.552 mg Carotene, 0.030-0.038 mg Thiamine, 0.019-0.03 mg Riboflavin and 0.294-0.380 mg Niacin per 100 g of fruit pulp. The wood of the *Averrhoa carambola* tree is often used in building furniture and used for construction (Morton, 1987). Due to the acidic properties of the *Averrhoa carambola,* its juice can be utilized to clean and polish metallic surfaces, as it dissolves rust and tarnish. On the other hand, the juice can also bleach rust stains from white clothing (Thulaja, 2016; Morton, 1987). Indisputably, fruits are very important in our daily diet for various health benefits. However, some fruits may contain high amounts of unique secondary metabolites, which are hazardous to our health. Star fruit plants are cultivated commercially in tropical countries for their fruits. This fruit have several medicinal properties; hence, it is used as medicine for many years in Ayurvedic treatments. Star fruits contain various antioxidants, which are considered medicinally important and beneficial for the health.

Health Benefits of Star Fruit

The fruit along with its waxy peel provides a good amount of dietary fiber. Fiber helps prevent absorption of dietary LDL-cholesterol in the gut. The dietary fibers also helps to protect the mucous membrane of the colon from exposure to toxic substances by binding to cancer-causing chemicals in the colon. In general, consumption of fruits rich in vitamin-C like Carambola helps the human body develop resistance against infectious agents and scavenge harmful, pro-inflammatory free radicals from the body. The Chinese use it to cure cough and cold and believe that it lowers blood pressure. Star fruit is rich in antioxidant phytonutrient *polyphenolic flavonoids.* Some of the important flavonoids present are quercetin, epicatechin, and gallic acid. Total polyphenol contents (Folin assay) in this fruit is 143-mg/100 g. Altogether, these compounds help to protect from deleterious effects of oxygen-derived free radicals by warding them off the body. Besides, it is a good source of B-complex vitamins such as folates, riboflavin, and pyridoxine (vitamin B-6). Apart from this it's a good source of minerals and electrolytes like potassium, phosphorus, zinc and iron. In India, the ripe fruit is administered to halt hemorrhages and to relieve bleeding hemorrhoids; and the dried fruit or the juice may be taken to counteract fevers. A conserve of the fruit is said to allay biliousness and diarrhea and to relieve a "hangover" from excessive indulgence in alcohol. A salve made of the fruit is employed to relieve eye afflictions. In Brazil, the Carambola is recommended as a diuretic in kidney and bladder complaints, and is believed to have a beneficial effect in the treatment of eczema. In Chinese Materia Medica it is stated, "Its action is to quench thirst, to increase the salivary secretion, and hence to allay fever." A decoction of combined fruit and leaves is drunk to overcome vomiting. The flowers are given as a vermifuge. In Southeast Asia, the flowers are rubbed on the dermatitis caused by lacquer derived from Rhus verniciflua Stokes. Burkill says that a preparation of the inner bark, with sandalwood and *Alyxia* sp., is applied on prickly heat. The roots, with sugar, are considered as an antidote for poison. Hydrocyanic acid has been detected in the leaves, stems and roots. A decoction of the crushed seeds acts as a galactagogue and ernmenagogue and is mildly intoxicating. The powdered seeds serve as a sedative in cases of asthma and colic (Morton, 1987).

Star fruit contains approximately 60 per cent of cellulose, 27 per cent of hemicelluloses and 13 per cent of pectin. It indicates that Star fruit is indeed rich in insoluble fibres fractions (IFF). The insoluble fibres have the ability to retain water more than cellulose; thus called as 'water insoluble fibre fractions' or WIFF, which lower the blood glucose by slowing down the absorption of carbohydrate in our body. In spite of this Consuming the fruit-juice together with the fibres (called as smoothie) does help in removing lipids through the excrement, and thus lowering the risks of cardiovascular diseases. It has also been reported that Star fruit extracts do have selective anti-brain tumour activity. Fibers also facilitate in lowering the total cholesterol level in the body by promoting hypo-glycaemic effect. Anti-inflammatory activity of Star fruit extracts help in lowering the skin inflammatory condition. Researchers induced a skin inflammatory condition akin to eczema using croton oil on a mice model. When ethanolic extracts of Star-fruit plant leaves were applied on the skin, it resulted in reduced inflammation and gradually reduced eczema in the mice. Traditionally, Star fruits are used to relieve stomach discomfort or any ulcer-like disorders. Extracts of Star-fruit plant leaves have anti-ulcerogenic properties. The extracts contain terpenoids (diterpenes and triterpenes), flavonoids and mucilage, which are known to have the anti-ulcer activity. The mucilage provides a lining to the gastro-intestinal mucosa, thus helping to avoid damages due to gastritis.

Food Uses

Ripe Carambolas are eaten out-of-hand, sliced and served in salads or used as garnish on avocado or seafood. They are also cooked in puddings, tarts, stews and curries. In Malaya, they are often stewed with sugar and cloves, alone or combined with apples. The Chinese cook Carambolas with fish and Thais boil the sliced green fruit with shrimp. Slightly under ripe fruits are salted, pickled or made into jam or other preserves. In Mainland China and in Taiwan, Carambolas are sliced lengthwise and canned in syrup for export. In Queensland, the sweeter type is cooked green as a vegetable. Cross-sections may be covered with honey, allowed to stand overnight, and then cooked briefly and, put into sterilized jars. Some cooks add raisins to give the product more character. A relish may be made of chopped unripe fruits combined with horseradish, celery, vinegar, seasonings and spices. Indian experimenters boiled horizontal slices with $3/4^{th}$ of their weight in sugar until very thick, with a Brix of 68^0. They found that the skin became very tough, the flavor was not distinctive, and the jam was rated as only fair. Sour fruits, pricked to permit absorption of sugar and cooked in syrup, at first 33^0 Brix, later 72^0, made an acceptable candied product though the skin was still tough. The ripe fruits are sometimes dried in Jamaica. Leaves are bound on the temples to soothe headache. Crushed leaves and shoots are poulticed on the eruptions of chicken pox, also on ringworm. Carambola juice is served as a cooling beverage. In Hawaii, the juice of sour fruits is mixed with gelatin, sugar, lemon juice and boiling water to make sherbet. Filipinos often use the juice as a seasoning. The juice is bottled in India, either with added citric acid (1 per cent by weight) and 0.05 per cent potassium meta-bisulphite (KMS), or merely sterilizing the filled bottles for ½ hour in boiling water. Star fruit and its juice are often recommended in many folk medicines in Brazil as a diuretic (to increase urine output), expectorant, and to suppress a cough. To make jelly, it is necessary to use unripe "sweet" types or ripe sour types and to add commercial pectin or some other fruit rich in pectin such as green papaya,

Food Value Per 100 g of Edible Portion of Carambola

Parameters	*Values*
Calories	35.7 Kcal
Moisture	89.0-91.0 g
Protein	0.38 g
Fat	0.08 g
Carbohydrates	9.38 g
Fiber	0.80-0.90 g
Ash	0.26-0.40 g
Calcium	4.4-6.0 mg
Phosphorus	15.5-21.0 mg
Iron	0.32-1.65 mg
Carotene	0.003-0.552 mg
Thiamine	0.03-0.038 mg
Riboflavin	0.019-0.03 mg
Niacin	0.294-0.38 mg
Ascorbic Acid	26.0-53.1 mg
Tryptophan	3.0 mg
Methionine	2 mg
Lysine	26 mg

together with lemon or lime juice. The flowers are acid and are added to salads in Java; also, they are made into preserves in India. The leaves have been eaten as a substitute for sorrel. It is often consumed fresh and also processed into jam, jelly, sweets, fresh juice and cordial concentrate.

Other Uses

The acid types of Carambola have been used to clean and polish metal, especially brass, as they dissolve tarnish and rust. The juice will also bleach rust stains from white cloth. The fresh leaves are used as animal fodder. The bark is used as a soap substitute in some areas. The mucilaginous extract of the bark is used in clarifying sugar and jaggery. Fiber extracted from the bark is used for rope making. The wood is yellowish-white, fine-grained, strong, and flexible. The pruned branches are used for basket making, support sticks and fuel wood purpose. The flower contains grewinol, a long chain keto alcohol (Laxmi and Chauhan, 1976). The seed of phalsa contains 5 percent oil, which is bright yellow in colour and contains 65 per cent linoleic acid, 13.5 per cent oleic acid and 11 per cent stearic acid (Morton, 1987). Unripe fruits are used in place of a conventional mordant in dyeing. Star fruit is one of the plant sources that contain the highest concentration of oxalic acid; 100 g of fresh fruit contains 50,000-95,800 ppm of oxalic acid, which treat as anti-nutrient and finally it predisposes to a condition known as oxaluria, a condition, which may lead to the formation of oxalate kidney stones. In some people with impaired kidney function, its consumption would result in renal failure and may cause death. Individuals with known kidney disease should, therefore, encourage for avoid eating star fruit.

Toxicological Effects

Star fruits do possess many magnificent properties. However, this fruit also poses threat to health as it exudes toxic effects in high uremic patients or patients with chronic renal disease due to its high oxalate content. Patients with renal disease are unable to secrete toxic substances out of their body efficiently; as a result of it, they are affected adversely by the oxalates. The first toxicological effect was demonstrated on mice model by Muir and Lam. Variable dosages of the fruit extracts were prepared and injected into the mice through intra-peritoneal injection, and fruit extracts exceeding 8 g/kg provoked convulsions and death in the mice. Further analysis of the test reports showed that Star-fruit juice with oxalate content was responsible for the death of rats. Chronic renal failure patients had high mortality rate after consuming the Star fruits. It was noted that these patients had symptoms of hiccups, mental confusions or disturbance in consciousness and vomiting before succumbing to death. Reports also suggest that uraemic patients experienced nephrotoxic and neurotoxic effects when they consumed Star fruit. Most of the patients were able to recover after immediate hemodialysis, which spanned for weeks but some experienced total renal failure, causing death. Even though Star fruits have many documented nutritional and medicinal benefits; but, due to the oxalate and caramboxin content in the fruits, it is considered toxic to patients experiencing renal problems.

However, the negative part of this fruit is that it produces oxalic acid and caramboxin, which are toxic to uremic patients. It can cause death if consumed insufficient quantity by those experiencing renal failure. Thus, more public awareness about oxalate poisoning on uremic patients should be promoted. It will help to avoid adverse side effect of Star fruits in high uremic patients. It is very important that the public is well educated on the benefits as well as the hazardous effects of the Star fruits (Muthu *et al.,* 2016).

Harvesting and Yields

The stage of maturity at harvest of the fruit influences the taste considerably. Since Carambola is a non-climacteric in respiratory behaviour (Lam and Wan, 1983), fructose and reducing sugar levels do not increase after harvest. The unripe fruit is light yellow and green at the edges whereas the fully ripen fruits has a deeper yellow colour with an orange tint (Gross *et al.,* 1983). Fruits of most cultivars should be picked at near full colour development to ensure maximum available sweetness. Wilson *et al.* (1985) identified forty-one volatile components in ripe fruit of cv. Golden Star out of which the most abundant one was methyl anthranilate, which has a distinct grape like aroma. The strong fruity aroma was mainly due to the esters and ketones. The Carambola tree normally begins to flower a few months after planting. There are 2-main fruiting seasons and 2 small seasons in a year. The two main fruiting seasons are April - June and October - December of each year. The fruits can be harvested after 45-90 days after flower anthesis and this depends very much on the weather and the clones. During rainy days, the fruits will ripen 2-3 days earlier compared to hot weather. It is advisable to allow the plant to reach the age of two years to bear fruits. Production will increase from year to year until 9 years and above where the production stabiles (about 1200 fruits/tree). The average weight of the Carambola fruit is between 17-66 tons/ha for the period of 20 years.

Packaging

Carambolas should be packed in strong, well-ventilated containers that have a smooth inner surface to protect against abrasion injury to the delicate fruit ribs. Wooden containers lined with newspaper, a thin cloth, or soft padding, are appropriate for the domestic market. Durable plastic crates are also acceptable. Synthetic or mesh sacks should not be used as they provide little or no protection to the fruit. Carambolas packed for export are typically put in well-ventilated fiberboard cartons of either 3 kg or 9 kg weight size, depending on market destination. The fruit should be oriented in a vertical position, with the stem end resting on an insert of foam padding. This minimizes damage to the rib edges (Anonymous, 2004).

Storage

According to Rathod *et al.* (2011) stated that the Carambola fruits stored in 200 gauge HDPE polyethylene bag with 1 per cent ventilation stored at refrigerated temperature conditions and pretreatment with 2 per cent $CaCl_2$ showed best result, but even freshly harvested fruits could be maintained by suitable plastic film packaging for at least 1 week at 20^0 C, the most important factor being control of moisture (and weight) loss. Cool storage at 5^0 - 10^0 C would extend the postharvest life of the fruit to at least 3 weeks (Brown and Wong, 1985). According to Kenny and Hull (1986) harvested mature Carambolas cultivars Fwang Tung and Arkin were harvested at colour break stage and placed into storage at 40^0, 45^0, 50^0 or 70^0 F for 8 weeks. All the samples stored with and without sealed polyethylene bags. Storage at 45^0 F for 28 days without using polyethylene bags did not significantly influence quality. While, cold damage was observed for both cultivars when stored at 40 ^{0}F. The cultivar Arkin and Golden Star was found better condition at 5 ^{0}C temperature with better appearance, less weight loss and less reduction in levels of soluble sugars (glucose, fructose and sucrose), and organic acid (oxalic and malic) as compare to high temperature storage (Campbell *et al.,* 1987). The phalsa is used for preparation of Phalsa juice (sarbat) and Phalsa squash. Phalsa juice ferments very rapidly and preservative such as sodium benzoate must be used for longer storage of the juice (Anand, 1960).

Value Added Products

Jam

Formulations of Fruit Jam (1000 g)

Ingredients	*Amount*
Pulp (g)	450
Sugar (g)	550
Citric acid (g)	5
Pectin (g)	5
KMS (ppm)	300
Water (ml)	250

Source: Monolisa ***et al.,*** 2014.

Jam Preparation In order to ensure the proper production of jam there are laid down procedures that have to be followed:

- ☆ Choose the fruits, which are firm and not over ripe mature or ripe since it contains more pectin, as this will make it easier for the jam to set. Unripe fruit does not have the same degree of flavor extremely delicious and acceptable jam needs.

- Wash fruit under clean running water or it can be washed with salt solution to get rid of any germs and grate them into pieces with gloves being worn.
- Weighed sugar (550 g) and add (250 ml) water to the sugar.
- Put sugar and water mixture on fire. The sugar must dissolve well before allowing the mixture to come to a boiling point. Allow boiling to a light gold caramel color.
- Add the fruit pulp (450 g) to the sugar mixture and allow cooking under gentle heat for 20 minutes. Bring the jam to a boiling point as quickly as possible to ensure that the jam reaches a point where it will solidify.
- Remove jam from fire and perform cold test. To know the end point of the jam, test for the setting point of the jam. A well-set jam transforms into a concentrated, thickened, viscous, uniformed, homogeneous clear form.
- Pour jam into clean sterilized bottles and cover with a tight lid.
- Jam must cool a little before applying the lid. Make sure no content of the jam gets on the rim of the jar. It can greatly add to friction after the lid is on. It actually makes the whole process unreliable. The jam might prevent a seal from forming or it might harden and make the lid difficult to remove. Always the rim of the jar must be clean before using it to cover the jam.
- Store in a cool, dark and dry place between 50-70^0 F temp (Egan, 2006).

Faults in Jam Preparation Though there are no strict guidelines regarding the production of jams, if the right procedures are not followed then the problems may create. The following are some of the mistakes or faults that are sometimes observed in jam preparation.

There are some problems occurs like the jam being too softened, due to a disproportionate amount of sugar, acid and fruit. A general recipe that is mostly used is adding one part under-ripe fruit to two parts fully ripe fruit so that the best gel and flavour can be formed. The USDA canning guide recommends at least one fourth of the fruit should be under-ripe. A large batch of the mixture, which can cause the jam to be under cooked, will also result in a jam, which is too soft. Another problem is jam that weeps. This happens as a result of a large amount of acid in the fruit or when too much acid from lemon juice is added. Pouring the jam too slowly into the jar can allow air bubbles to form and leaving the mixture after cooking to stand for a long time before pouring it into a jar can also make the jam cloudy. When the jam is not properly sealed or airtight, mould can grow on it. Mould growth on jam should be avoided because consumers will not buy mouldy jam and also mould growth will cause severe food poisoning (Singh-Ackbarali and Maharaj, 2014). Commercial pectin can be added to any fruit, including fruits that have high pectin content. Adding too much pectin will result in a tough, rubbery consistency, which is difficult to spread but the proper recipe guide step by step that comes with the commercial pectin will help annihilate this difficulty (Bastin, 2004). According to Monolisa *et al.* (2014) consumers can consume the jam without changing quality facts starting 0 day to 3 months at any storage temperature and the product can be store for 6 months.

Squash

Formulation of Carambola fruit Squash

Ingredients	*Amount*
Juice (g)	250
Sugar (g)	372
Citric acid (g)	11.50
KMS (g)	0.52
Water (ml)	365.54

Source: Monolisa ***et al.,*** 2014.

Sugar, citric acid, juice and KMS were weighted as required separately. Water is also measured according to calculation. Then, sugar and acid mixed in water (a small quantity of water to be taken out from the measured quantity to dissolve KMS later on). After that, boiled the mixture and strained through filter. Then, cool the syrup. The above syrup was mixed with the juice. Then KMS was dissolved in aforesaid quantity of water and added to the squash. Finally, poured into pasteurized bottles and sealed. According to Monolisa *et al.* (2014) consumers can consume the squash without changing quality facts starting 0 day to 3 months at any storage temperature and the product can be stored for 6 months.

REFERENCES

Anand, J. C. (1960). Efficacy of sodium benzoate to control yeast fermentation in phalsa (*Grewia asiatica* L.) juice. *Indian Journal Horticulture.* 17: 138–141.

Anonymous (2004). Carambola (Five Fingers). Postharvest Handling Technical Bulletin. Ministry of Fisheries, Crops and Livestock, New Guyana Marketing Corporation, National Agricultural Research Institute Technical Bulletin No. 30, pp. 1-10.

Bastin, S. (2004). The Science of jam and jelly making. Cooperative Extension Service. University of Kentucky. College of Agriculture.

Bose, T. K., Mitra, S. K. and Sanyal, D. (2002). Fruit: tropical and subtropical. Nayaudyog, III Revised edition. 2: 78-596.

Brown, B. I. and Wong, L. S. (1985). Postharvest handling technology for new tropical and sub-tropical fruit crops, *In*: Postharvest storage of Carambola (*Averrhoa carambola* L.). Queensland Department of Primary Industries, Hamilton, Brisbane, pp. 4.

Campbell, C. A., Huber, D. J. and Koch, K. E. (1987). Postharvest response of Carambolas to storage at low temperatures. Proc. Fla, State Hort. Soc., 100: 272-275.

Chang, C. T., Chen, Y. C., Fang, J. T. and Huang, C. C. (2002). Star fruit (*Averrhoa carambola*) intoxication: an important cause of consciousness disturbance in patients with failure. *Pubmed* 24 (3): 379-82.

Eagan, S. (2006). Small Scale Production of Food Preserves. Fact Sheet No. 16, Agriculture and Food Development Authority.

Gross, J., Ikon, R. and Eckhardt, G. (1983). *Phytochemm.* 22: 1479-53.

Kanwar, A. and Budhwar, S. (2018). Utilization of star fruit (*Averrhoa carambola*) into value added products and their storage stability. *International Journal of Food Science and Nutrition.* 3(1): 77-80.

Kenney, P. and Hull, L. (1986). Effect of storage condition on Carambola quality. *Proc. Fla, State Hort. Soc.,* 99: 222-224.

Lakshmi, V. and Chauhan. J. S. (1976). Grewinol, a keto-alcohol from the flowers of *Grewia asiatica. Lloydia* 39: 372–374.

Lam, P. F. and Wan, C. K. (1983). Climacteric nature of the Carambola (*Averrhoa carambola* L.) Fruit. *Pertanika.* 6(3): 44-47.

Monalisa, K., Islam, M. Z., Asif-Ul-Alam, S. M. and Hoque, M. M. (2014). Valorization and Storage Stability Assessment of Underutilized fruit Carambola (*Averrhoa carambola*) in Bangladesh. American Journal of Food Science and Technology. 2(4): 134-138. doi: 10.12691/ajfst-2-4-5.

Morton, J. (1987). Carambola. *In: Fruits of warm climates.* Julia F. Morton, Miami, FL. p. 125–128.

Muthu, N., Lee, Su Yin., Phua, Kia Kien and Bhore, S. J. (2016). Nutritional, medicinal and toxicological attributes of Star Fruits (*Averrhoa carambola* L.): A Review. *Bioinformation.* 12 (12): 420-424.

Rathod, A., Shoba, H. and Chidanand, D. V. (2011). A Study on Shelf life Extension of Carambola Fruits. *International Journal of Scientific and Engineering Research.* 2(9): 1-5.

Singh-Ackbarali, D., and Maharaj, R. (2014). Sensory evaluation as a tool in determining acceptability of innovative products developed by undergraduate students in food science and technology at the University of Trinidad and Tobago. Journal of Curriculum and Teaching. 3(1). [Online] Available: http://www.sciedu.ca/jct https://www.hort.purdue.edu/newcrop/morton/Carambola.html 4/6

Srivastava, K. K. and Rajput, C. B. S. (2003). Genetic diversity in Carambola (*Averrhoa carambola* L.). *Indian Horticulture.* Oct – Dec., pp. 2.

Thulaja, N. R. (2016). Star fruit. National Library Board - Singapore. Retrieved from http://eresources.nlb.gov.sg/infopedia/articles/SIP_173_2005-01-06.html

USDA (2005). US Department of Health and Human Services. Dietary Guidelines for Americans. Washington, DC: USDA.

Wilson, C. W., Shaw, Knight, P. E., R. J. Jr., Nagy, S. and Klim, M. (1985). Volatile constituents of Carambola (*Averrhoa carambola* L.). *Journal of Agricultural and Food Chemistry*. 33: 199-201.

5

Fig

Scientific Name: *Ficus carica* L.

Family: Moraceae

The Fig (*Ficus carica* L.), one of the first cultivated trees in the world, is grown in many parts of the world with moderate climates (Solomon *et al.,* 2006). It's a deciduous fruit tree belongs to a member of the Mulberry family Moraceae which includes Mulberries and many other tropical trees such as Breadfruit and Jackfruit. The common Fig (*Ficus carica* L.) is known a traditional fruit in Western Asia. Its historical origin is presumably southern Arabia (Stover *et al.,* 2007) or the eastern part of the Mediterranean area including Turkey and Iran where wild forms of Fig trees can be observed. The acreage of Fig in the world reached about 415325 hectare, with production of 1043989 MT. Egypt is considered as the second producing country in the world whereas cultivated area of Fig reached 28479 hectare with 165483 tons of the total production. The five major Fig producing country in the world is Turkey, Egypt, Iran, Algeria and Morocco which accounts 64 per cent of the total world production of Fig (FAO, 2007 and 2011; Agriculture Dept. MH. 2007-08). Fig is an important fruit, which has been under cultivation since ancient time. Fig was not cultivated as commercial crop in India till date, but now a days its cultivated in some extent in basically in three major states like Maharashrta, Karnataka and Andhra Pradesh with an area of 2899 ha and production was 13930 m. tonnes.

The Fig is a type of climacteric fruit; Figs are a nutritious fruit rich in fiber, potassium, calcium, and iron with higher levels than other common fruits such as bananas, grapes, oranges, strawberries, and apples (Chessa, 1997; Michailides, 2003). Figs are having free of sodium, fat and cholesterol like other fruits. Additionally, Figs are an important source of vitamins, amino acids, and antioxidants (Solomon *et al.,* 2006). Botanically, the Fig cannot be called a fruit. It is a syconium, which means a part of the stem extends into a sac, with the flowers growing internally. They are known by various names in local languages. They are called 'Anjeer' in Hindi, 'AthiPallu' in Telugu, 'Atti Pazham' in Tamil and Malayalam, 'Anjura' in Kannada, and 'Dumoor' in Bengali. It is a scion of the fruit and known as a false fruit or multiple fruit. The small orifice (ostiole) visible on the middle of the fruit is a narrow passage, which allows a specialized Fig wasp Blasto phagapsenes to enter the fruit and pollinate the flower, where after the fruit grows seeds. The edible fruit consists of the mature syconium containing numerous one-seeded fruits (druplets). The fruit is become 3–5 centimetres (1.2–2.0 inch) long with a green skin, sometimes ripening towards purple or brown. *Ficus carica* has milky sap (laticifer), the sap of the Fig's green parts is an irritant to human skin (McMahon *et al.,* 2002).

Types of Figs

There are five common varieties of Figs. Each type differs subtly in flavor and sweetness. They are:

1. **Black Mission–** Black Mission Figs are blackish-purple outside and pink inside. They are incredibly sweet and even ooze out the syrup. They are perfect to eat as a dessert or mix in cake or cookie recipes to increase the flavor.
2. **Kadota–** Kadotas are green with purple flesh. They are the least sweet among all the varieties of Figs. They are excellent to eat raw and also taste good if heated with a pinch of salt.
3. **Calimyrna–** Calimyrna Figs are greenish-yellow on the outside and amber on the inside. They are larger when compared to another type of Figs and have a unique and strong nut flavor.
4. **Brown Turkey–** The Brown Turkey Figs have purple skin and red flesh. Their flavor is mild and less sweet than the other type of Figs. They work well in salads.
5. **Adriatic–** Adriatic Figs have light green skin and are pink on the inside. These Figs are often used to make Fig bars. They are also called white Figs, as they are very light colored. They are extremely sweet and can be enjoyed as a simple fruit dessert.

Importance and Use

But day by day, more and more land is being brought under cultivation of Fig. Farmers who engaged in the fostering of this fruit were being benefited to a large extend. Fig contains minerals such as iron, copper and calcium in a large quantity and plenty of various vitamins. Food items like dry Fig, sweetmeat, Fig milkshake *etc.* are made from Fig. Due to all these reasons, Fig has become a significant fruit. As Fig blooms rapidly after cultivation, it becomes possible to bloom the Fig in the consecutive year, which facilitates the immediate earnings. Figs also include variety of antioxidants and omega fatty acids and Figs have a great importance for nutritional quality due to being important source of carbohydrates. Various parts of the plant like bark, leaves, tender shoots, fruits, seeds, and latex are medicinally important. The Fig is a very nourishing food and used in industrial products. Figs are one of the highest plant sources of calcium and fiber. They are good source of flavonoids and polyphenols and some bioactive compounds such as arabinose, β-amyrins, β-carotines, glycosides, β-setosterols and xanthotoxol (Gilani *et al.,* 2008). They contain essential amino acids and are rich in vitamins B_1, B_2 and C and minerals. The dried Figs produced a significant increase in plasma antioxidant capacity and also used in various disorders such as gastrointestinal respiratory, inflammatory, cardiovascular disorders, ulcerative diseases and cancers (Mcgovern, 2002). It has been mentioned in 2900 B.C. for its medicinal use. The health benefits of Fig or anjeer include its use as a treatment for sexual dysfunction, constipation, indigestion, piles, diabetes, cough, bronchitis and asthma. It is also used as a quick and healthy way to gain weight back after suffering through an illness (Doyle *et al.,* 2006). In traditional medicine, the roots are used in treatment of leucoderma and ringworms and it's also fruits, which are sweet, have antipyretic, purgative, aphrodisiac properties and have shown to be useful in inflammations and paralysis (Kirtikar and Basu, 1996). *F. carica* has been reported to include antioxidant, antiviral, antibacterial, antipyretic, antifungal (Houda *et al.,* 2010), hypoglycemic, cancer suppressive, hypo-triglyceridaemic, and anthelmintic effects (Jeong, 2005).

There are several other uses:

- ✰ Chew 2-3 tender Fig leaves and gargle with water to overcome bad breath and ulcer. Fig is an orthodox remedy to increase sexual weakness. Soak 4-5 Figs in milk overnight and eat in the morning to overcome weakness (Franks, J. 2013). Boil 6 s in a cup of water and consume its daily to remove kidney stone.
- ✰ Apply mashed fresh Figs all over your face and let it dry for 15-20 minutes. This remedy is useful for curing acne. Also milky juice of Fig stem and leaves can be applied daily several times to cure warts. They also have anti-ageing properties.
- ✰ Since Figs are high in potassium, they help reduce insulin. Daily, consume Fig seeds with one teaspoon with honey to control diabetes. Fig leaves can be consumed for anti-diabetic properties.
- ✰ Figs have good amount of calcium in it which help in strengthen bones.
- ✰ Early stages of chicken pox can be treated with Figs.
- ✰ Figs include good amount of dietary fibers, which are helpful for reducing body weight (Howard and Patten, 1960).
- ✰ Mix dries Fig and honey in water to get rid of sore throat.
- ✰ Consume 2-3 soaked dried Figs with a tablespoon of honey. Use this continuously for a month every morning to relive constipation.

- Figs are rich in important nutrients like Vitamin B, C, phosphorus, potassium and minerals like calcium and magnesium which are essential for boosting and rejuvenating the skin's health. The high omega 3 fatty acids in Figs keep the skin well-moisturized and conditioned from within (Gordon, 2007).
- A Fig is not just a delicious and healthy fruit but is a great natural ingredient for skin care. Applying baked Fig directly on the skin brings down various forms of skin inflammation like boils and abscess. It also cures minor zits and pimples. Apply Fig juice on your feet to cure corns.
- Figs are loaded with anti-oxidants and dietary fiber which make it an excellent natural laxative for the body. It helps to flush out toxins and waste from the body to prevent skin conditions like acne and psoriasis and gives you healthy and glowing skin (Franks, 2013).
- Application of Fig paste on the face helps to transport important nutrients into the dermal layers of the skin to replenish it. Take some fresh Figs and process them in a mixer with 1 tablespoon of yoghurt to make a fine paste. Apply this all over the face and massage you skin gently with a few minutes. Leave it for 15 minutes and then wash off with lukewarm water. If you cannot find fresh Figs then soak some dried Figs overnight and blend them to form a smooth paste (Rolek, 2015).
- You can also prepare a natural, homemade scrub using Fig paste. Mix 1 tablespoon of powdered sugar and 2 tablespoon of fresh orange juice with 2 tablespoon Fig paste. Add a few drops of olive oil and use this scrub on both your face and body weekly to get smooth and soft skin. The enzymes present in Figs will remove the dead skin cells and the sugar will mildly polish the skin (Weigl, 2015).
- Figs contain a high amount of Vitamin C which helps to lighten and even out the skin tone. Prepare a Fig paste by mixing powdered oatmeal and dried ginger powder. Add a few drops of bergamot essential oil and stir to form a smooth paste. Use this face pack twice a week to get even-toned skin.
- Not just the flesh, the peels of Figs too contain active enzyme which act as an exfoliator to clear the dead cell debris from the skin. These all need to do is wash and clean the peels, and scrub the skin with the inside of the peel for flawless and smooth skin (Brothwell and Brothwell, 1998).
- Figs have amazing moisturizing properties as they contain around 3/4 cups of water. It is often used as a moisturizer on the skin to restore its suppleness. Apply some Fig paste on your lips to alleviate cracked lips. Application of Fig paste on the face helps to tighten the pores and control excess sebum secretion (Tannahill, 1998).

Health Benefits of Figs

1) Enhance Digestive Health

Figs relieve constipation and improve the overall digestive health. Fiber is great for digestion, and Figs are loaded with dietary fiber, which aids healthy bowel movement and relieves constipation. It adds bulk to the stools and promotes their smooth passage through the body. The fiber in Figs also treats diarrhea and soothes the entire digestive system.

2. Improve Heart Health

Figs reduce the triglyceride levels in the blood and contribute to improving heart health. Tri-glycerides are fat particles in the blood that are a leading cause of heart diseases. Also, the antioxidants in Figs get rid of the free radicals in the body, which block the coronary arteries and cause coronary heart disease. Figs also contain phenols and omega-3 and omega-6 fatty acids that decrease the risk of heart disease.

3. Lower Cholesterol

Figs contain pectin, a soluble fiber that is known to reduce cholesterol levels. The fiber in Figs clears the excess cholesterol in the digestive system and carries it to the bowels to eliminate it. Figs also contain vitamin B-6 that is responsible for producing serotonin. This serotonin boosts mood and lowers cholesterol. Dry Figs reduce the overall cholesterol as they contain omega-3 and omega-6 fatty acids and phyto-sterols that decrease the natural cholesterol synthesis in the body.

4. Prevent Colon Cancer

Regular consumption of Figs can lower the risk of colon cancer. The fiber in Figs helps to eliminate the waste in the body quickly, which works well for the prevention of colon cancer. The numerous seeds in Figs contain high levels of mucin that collects wastes and mucus in the colon and flushes them out.

5. Cure Anemia

Lack of iron in the body can cause iron-deficiency anemia. Dried Figs contain iron, which is a key component of hemoglobin. Consuming dried Figs was found to improve the hemoglobin levels in the blood.

6. Strengthen Bones

Figs contain calcium, potassium, and magnesium, all of which aid bone health. Figs improve bone density and decrease the breakdown of bones. Figs contain potassium that counteracts the increased urinary calcium loss caused by high-salt diets. This prevents bones from thinning out.

7. Rich In Antioxidants

Figs are the power house of antioxidants, and they neutralize the free radicals in the body and Fight diseases. The riper a Fig is, the more antioxidants it contains.

Figs are a rich source of phenolic antioxidants. The antioxidants in Figs enrich the lipoproteins in plasma and shield them from further oxidation.

8. Regulate High Blood Pressure

Studies show that including Figs in our daily diet helps to lower blood pressure. The fiber in Figs lowers the risk of high blood pressure whereas the potassium content of Figs helps to maintain it. Apart from potassium, the omega-3s and omega-6s in Figs also help in maintaining blood pressure.

9. Prevent Hypertension

When we consume less of potassium and more of sodium, it disturbs the sodium-potassium balance in our body, paving the way for hypertension. Figs help restore this balance, as they are rich in potassium.

10. Treat Asthma

Figs moisturize the mucous membrane and drain the phlegm, thereby relieving asthma symptoms. They also contain phytochemical compounds that Fight the free radicals, which otherwise trigger asthma.

11. Lower Sugar Levels In Diabetic Patients

Fig leaves have amazing properties that help regulate blood glucose levels. According to a study, including Fig leaves in the diet helped control the rise in blood sugar post a meal in insulin-dependent diabetics.

12. Prevent Macular Degeneration

Figs can help prevent macular degeneration, which is a leading cause of vision loss in older people as they contain a high amount of vitamin A.

13. Treat Piles

Dry Figs are the best to treat piles. The seeds in the Figs are the active agents that fight the piles.

14. Prevent Coronary Heart Disease

The antioxidants in Figs, as well as their blood pressure lowering properties, eliminate the free radicals in the body, which otherwise block the coronary arteries, leading to coronary heart disease. Presence of potassium, omega-3s, and omega-6s in Figs help in preventing heart attacks.

15. A Good Source of Energy

The carbohydrates and sugar present in Figs increase the percentage of energy in your body.

16. Boost the Immune System

Figs kill bacteria, viruses, and roundworms in your body, which can otherwise cause health issues. They contain nutrients like potassium and manganese that, along with the antioxidants, boost your immune system.

Physical Characteristics of Plant

- ✩ Deciduous tree from family-Moraceae.
- ✩ It is of 4m height and much more wide and tall.
- ✩ It has smooth stems with grey bark.
- ✩ Leaves are alternate with 3 to 5 round lobes, very rough to touch above and pubescent below.
- ✩ Flowers inside are pear shaped receptacle which later becomes fruit.
- ✩ Green fruit when unripe that matures into yellowish green or dark purple

Chemical Composition

- ✩ Amino acids (alanine, arginine, aspartame, cysteine, lysine, glycine, lipase, methionine)
- ✩ Enzymes (esterase, ficin)
- ✩ Carbohydrates (glucose, galactose)
- ✩ Vitamins (vitamin A, vitamin C, niacin)
- ✩ Linoleic acid, malicacid, oleic acid
- ✩ Minerals (potassium, phosphorus, magnesium, manganese, copper, calcium)

Composition of Fig

Nutrients	*Composition (per cent) in 50g of Dried Fig*	*Nutrients*	*Composition (per cent) in 50g Fresh Fig*
Protein	2.5 per cent	Fibre	15 per cent
Fat	<1 per cent	Magnesium	12 per cent
Carbohydrates	48 per cent	Potassium	8 per cent
Fibre	29 per cent	Carbohydrates	25 per cent
Potassium	14 per cent	Vitamin K	7 per cent
Vitamin K	14 per cent	Vitamin B6	7 per cent
Calcium	12 per cent		

Source: Isabel and Schafer, 1990.

Harvesting and Yield

Figs are easy to harvest from tree when it become ripe and avoid the bruising. Pull or cut the fruit gently from the stem, leaving some of the stem attached to the Fig to help delay fruit spoilage.

Keep the Figs in a shallow dish and do not pack them tightly on top of each other, as they bruise easily. Use caution when working above your head or on a ladder. If you have a tall tree, it's helpful to have an assistant while you pick.

Caution: Some people are allergic to Fig latex, the milky white sap that oozes from the leaves and branches, and from the stems of unripe Figs. The sap can cause itchy, painful dermatitis that can become worse when exposed to sunlight. If you are allergic to latex, be sure to wear long sleeves and gloves when harvesting of Figs.

When to Pick Figs

Wait until the Figs ripe to harvest. Figs are not continued to ripen after they are picked like many other fruits. It is time for harvesting of Figs when the fruit necks wilt and the fruits hang down.

If fruit pick too early, it will taste horrible; ripe fruit is sweet and delicious. As long as the fruit is still perpendicular to the stem, it is not ready to be picked. A perfectly ripe Fig will also emit its nectar at its peak and be soft to touch. It is always better to error on the side of picking a Fig that is slightly overripe than under ripe. The fruit will change the colour as it gets riper. Each Fig type has different colors and ripeness can vary from green to dark brown. Be sure to harvest in the morning on a partly cloudy day for best results.

Age of the Plant (Years)	*Yield/Tree (kg)*
3	3
4	6
5	9
6	12
7	15
8th year onwards	18

Source: Patten, 2001.

Fig Selection

The shelf life of fresh Figs is brief indeed. They must be picked ripe from the trees as they do not ripen well once picked. A very firm Fig is not ripe and will not properly ripen further. The harvesting season for fresh Figs is mid-June to mid-October. It will spoil within seven to ten days of harvesting because it is a highly perishable in nature. Select Figs that are clean and dry, with smooth, unbroken skin. The fruit should be soft and yielding to the touch, but not mushy. It can be also check the stage of maturity by smells, if it smells slightly sour; it has already begun to ferment. When Figs get beyond their prime, they begin to collapse inward and lose their round shape (Howard and Patten, 1960).

Storage

It is the fact that Figs are highly perishable in nature, which could not storage for long periods, and in order to expand the potential markets. So, most of the production is used for drying. Figs are climacteric type of fruits and are slightly sensitive to ethylene action on stimulating softening and decay severity, especially if kept at 5 °C or higher temperatures. However, low temperatures from 0 to 2 °C and a high relative humidity (90 to 95 per cent RH) are recommended for storage. Good quality of fresh Figs may be maintained for up to 4 weeks when kept at 0, 2.2, or 5 °C in atmospheres enriched with 15 per cent or 20 per cent CO_2. Fresh Figs are very sensitive to microbial spoilage, even in cold storage conditions; thus they must be preserved in some way. Use them immediately or store in a plastic bag in the coldest part of your refrigerator for up to two days. Figs can be frozen whole, sliced or peeled into a sealed container for ten to twelve months. Canned Figs will be stored for a year in good condition. Dried Figs can be stored in the original sealed package at room temperature for a month. It can be stored for longer duration in the refrigerator upto a year.

Turk *et al.* (1994) determined that fruits of *cv.* Bursa Siyahi could be stored for 4 weeks at 0 °C and 90 to 95 per cent RH with CO_2:O_2 ratio of 3.3 and 5.5. The shelf life of fruits could be extended by storing fruits at 0 °C atmospheres enriched with 20-25 per cent CO_2. Unwrapped fruits were not suitable for marketing after two days of storage. The high level of CO_2 in the storage atmosphere greatly reduce rates of respiration, ethylene production, decay incidence and significantly contributed to maintaining good quality in Calymirna and Mission Figure

The Indian meal moth, *Plodia interpunctella* that is a major pest in dry Fig storage, can be satisfactorily controlled by packaging Figs in plastic bags having a thickness of 0.24 mm after suitable temperature treatment. Hot air treatments at 60 °C for at least 7 hours or at 65 °C for at least 6 hours are effective in controlling pest infestation without affecting fruit quality (Rahemi and Zare, 2002).

Value Added Products

Preparation of Fig Pulp

Technical Flow-Chart for Processing of Fig Pulp

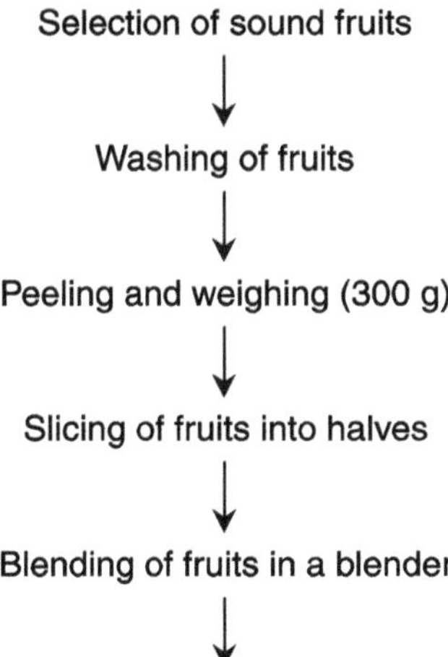

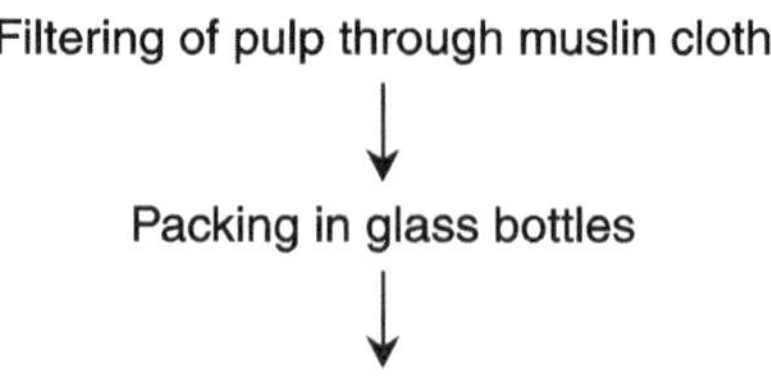

Source: Tanwar ***et al.,*** 2014

Preparation of Fig Jam

Sl.No.	*Ingredients*	*Quantity*	*Sl.No.*	*Ingredients*	*Quantity*
1	Fruit	1 kg	3	Citric acid	2.5 g
2	Sugar	1 kg			

Technical Flow-Chart for Processing of Fig Jam

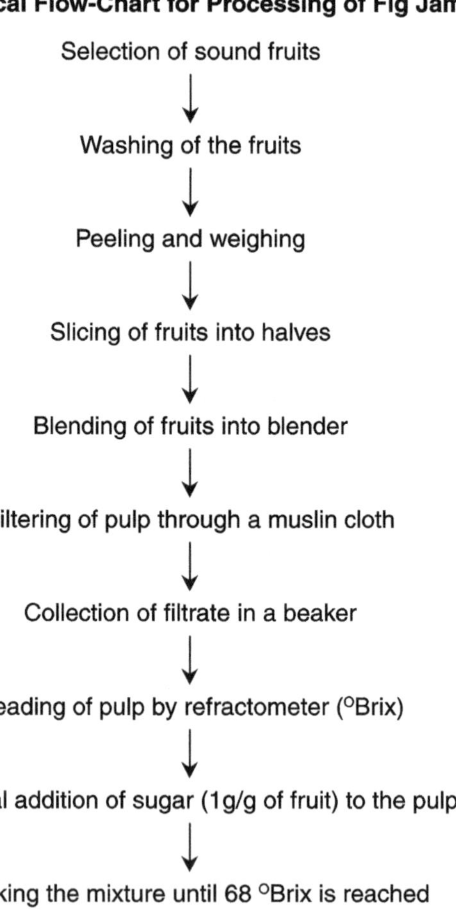

Preparation of Fig Nectar

Sl.No.	*Ingredients*	*Quantity*	*Sl.No.*	*Ingredients*	*Quantity*
1	Fruit	1 kg	3	Citric acid	2.5 g
2	Sugar	0.6 kg			

Technical Flow-Chart for Processing of Fig Nectar

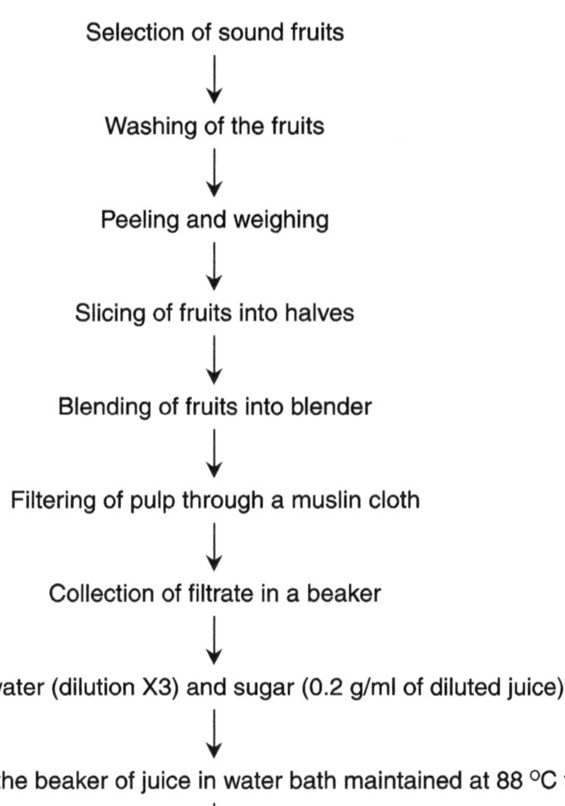

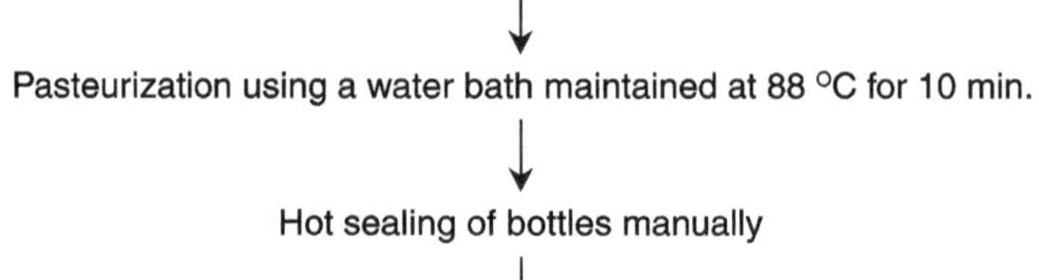

Source: Tanwar ***et al.***, 2014

Fig and Feta Gau Gee and Wontons

Sl.No.	*Ingredients*	*Quantity*	*Sl.No.*	*Ingredients*	*Quantity*
1	Fig (Ripe)	6 pcs	4	Garlic chopped	1 tsp
2	Cheese	124 g	5	Package wonton pepper	-
3	Fresh ground pepper	-			

Procedure

- ☆ Wash and cut off stem end of Figs.
- ☆ Put Figs, cheese, and garlic into food processor or blender and pulse slowly.
- ☆ Texture should be slightly lumpy and not liquid. Season with a pinch of fresh ground pepper.
- ☆ Spread about 1 teaspoon of the mixture onto a wonton and fold to desired shape.
- ☆ Dampen edges of the wonton so it sticks together.
- ☆ Deep-fry wontons until golden brown.
- ☆ Makes about 50 pieces.
- ☆ You can also add finely chopped fresh spinach and cooked rice or orzo pasta to the mixture if desired.
- ☆ You can also steam the wontons or form them into shumai.
- ☆ Serve with sweet and sour dipping sauce or spicy chili sauce (Tanis, 2013).

Fig Preserve

Sl.No.	*Ingredients*	*Quantity*	*Sl.No.*	*Ingredients*	*Quantity*
1	Baking soda	2 tsp	4	Sugar	1 ½ cup
2	Fresh Fig weithout stem	5 cups	5	Butter	5 tsp
3	Water	1 cup	6	Vanilla extract	1 tsp

Sl.No.	*Ingredients*	*Quantity*	*Sl.No.*	*Ingredients*	*Quantity*
4	Lemon juice	1 tsp	10	Ground cloves	½ tsp
5	Lemon (Thinly sliced into round)	1	11	Salt	As par taste
6	Grated fresh ginger	1 tsp	12	Ground cinnamon	1 ½ tsp

Procedure

- ☆ Dissolve the baking soda into 2 quarts of cool water and immerse the Figs in the treated water in a large bowl. Gently stir to wash the Figs, then drain off the water and rinse the Figs thoroughly with fresh cool water. Place the Figs into a large pot. Add 1 cup water, sugar, butter, vanilla extract, lemon, lemon juice, cinnamon, ginger, and cloves. Very gently stir the mixture to dissolve the sugar, keeping the Figs intact as much as possible.
- ☆ Bring the mixture to a boil over medium heat; reduce heat to a simmer, and cook until the Figs are golden brown and coated in syrup, about 1 hour. Stir gently a couple of times to keep the Figs from burning onto the bottom of the pot. Add a pinch of salt, if desired, to tame the sweetness.
- ☆ Sterilize the jars and lids in boiling water for at least 5 minutes. Pack the Figs into the hot, sterilized jars and top off with syrup, filling the jars to within 1/4 inch of the top. Run a knife or a thin spatula around the insides of the jars after they have been filled to remove any air bubbles. Wipe the rims of the jars with a moist paper towel to remove any food residue. Top with lids, and screw on rings.
- ☆ Place a rack in the bottom of a large stockpot and fill halfway with water. Bring to a boil over high heat, then carefully lower the jars into the pot using a holder. Leave a 2 inch space between the jars. Pour in more boiling water if necessary until the water level is at least 1 inch above the tops of the jars. Bring the water to a full boil, cover the pot and process for 15 minutes.

Orange, Fig and Gongorzola Salad

Sl.No.	*Ingredients*	*Quantity*	*Sl.No.*	*Ingredients*	*Quantity*
1	Lettuce chopped	2 heads	3	Gorgonzola cheese	½ cup
2	Oranges (peeled, pith removed and cut into segments)	2	4	Fresh Fig (cut into 1 inch cubes)	2

Procedure

- ☆ Combine lettuce, oranges, Gorgonzola cheese, and Figs in a large bowl. Drizzle dressing over salad and toss to coat.

Fig Cookies

Cookies were prepared by adding Fig powder in 0, 6, 12 and 18 per cent mixed with butter or vegetable fat, sugar, baking powder and beatened eggs and dough was prepared. The dough was cut in to rectangular shape and baked on greased pan in an oven at 160 ^{0}C for 15 min.

Sl.No.	*Ingredients*	*Quantity*	*Sl.No.*	*Ingredients*	*Quantity*
1	sugar	1 cup	6	Salt	½ tsp
2	Shortening	½ cup	7	Ground cloves	½ tsp
3	Egg	1	8	Fresh Fig	1 cup
4	All-purpose flour	2	9	Chopped walnut	½ cup
5	Baking Soda	1 tsp			

Procedure

- ☆ Preheat oven to 350 ^{0}F (180 ^{0}C).
- ☆ Cream sugar and shortening and add beaten egg.
- ☆ Sift dry ingredients and blend with creamed mixture. Fold in Figs and nuts.
- ☆ Drop by spoonfuls on greased sheet. Bake for 15 to 20 minutes.

Flow Sheet for Preparation of Fig Powder Enriched Cookies (Khapre *et al.,* 2015)

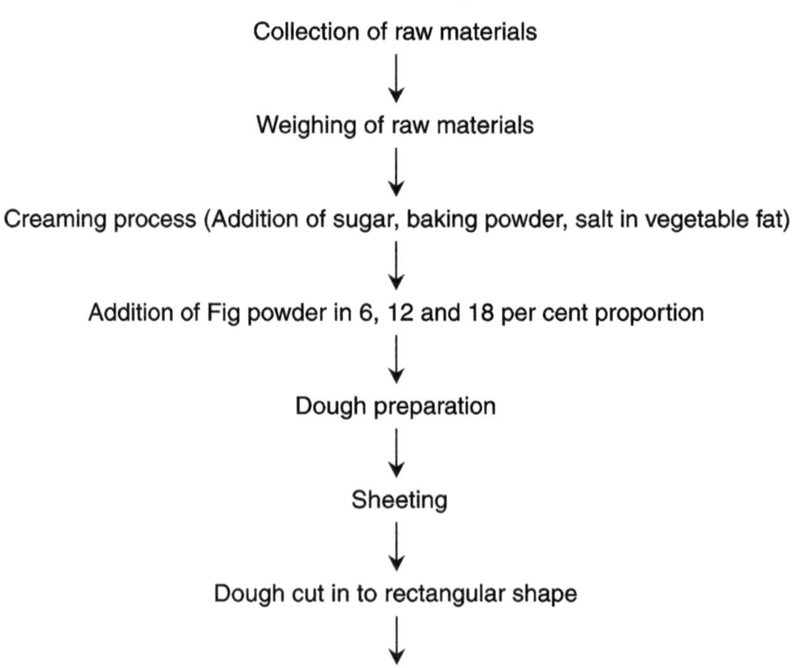

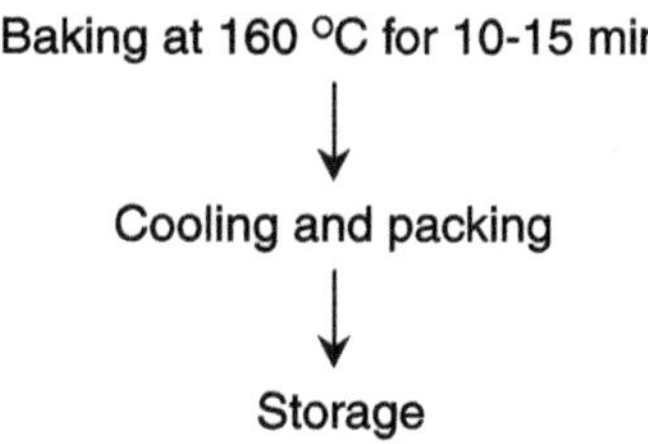

Khapre *et al.* (2015) reported 3.1 per cent dietary fiber in Fig cookies. The Fig cookie also contained 76 per cent total soluble solids. It was also reveals that the Fig cookie contained reducing and non-reducing sugar 53.1 and 22.9 per cent respectively. Fig cookie was rich in protein and contained 6.9 per cent protein. The ascorbic acid of cookie decreased as compared to fresh Figs and it was 3.8. mg/100 g. The value of potassium found in Fig cookie was 1100 mg/100g and therefore it is rich source of potassium.

Chocolate Fig Bread Pudding

Sl.No.	*Ingredients*	*Quantity*	*Sl.No.*	*Ingredients*	*Quantity*
1	White bread	4 cup	6	Egg	2
2	Fig chopped	1 cup	7	Granulated sugar	¼ cup
3	Coarsely chopped chocolate	½ cup	8	Vanilla	2 tsp
4	Milk	2 cups	9	Salt	¼ tsp
5	Butter	2 tsp			

Spirited Sauce

Sl.No.	*Ingredients*	*Quantity*	*Sl.No.*	*Ingredients*	*Quantity*
1	Sugar (powder)	1 ¼ cup	3	Egg	1
2	Butter	4 tsp	4	Dark rum/Brandy	¼ cup

Procedure

- ☆ Heat oven to 350 °F. Grease 1½ -quart baking dish. Combine bread cubes, Figs and chocolate in baking dish. Heat milk and butter until butter is melted (do not boil). Combine eggs, sugar, vanilla, and salt; blend well. Add hot milk mixture to egg mixture, stirring constantly. Pour over bread and Fig mixture. Let stand 5 minutes. Bake uncovered for 50 minutes or until knife inserted in centre comes out clean. If desired, baked pudding can be made several days ahead.
- ☆ Cover and refrigerate. Reheat in 325 °F oven for 15 minutes. Prepare Spirited Sauce. Serve warm over bread pudding.
- ☆ Spirited Sauce: In saucepan, combine all sauce ingredients. Cook over low heat, stirring constantly, about 3 minutes.

Fig Toffee

Toffee is one of the essential products largely consumed by children. The confectionary products due to its varied taste and flavor have a wide acceptance in children throughout the world. The conventional toffees are generally made from sugar, skim milk powder and other artificial colours and flavors (Mahalaskar *et al.,* 2012).

REFERENCES

Brothwell, D, and Brothwell, P. (1998). Food in Antiquity: A survey of the diet of early people. Johns Hopkins University Press, Baltimore and London. pp. 144–147.

Chessa, I. (1997). In: Mitra, S. (ed.). Postharvest physiology and storage of tropical and subtropical fruits. CAB International, Wallingford, UK. pp. 245–268.

Doyle, *et al.* (2006). USPTO Patent Full-Text and Image Database, Fig Tree Named 'Sequoia.

Food and Agriculture Organization (2007). Data archives. FAOSTAT. Available from: http://faostat.fao.org/site/408/DesktopDefault.aspx [Accessed 24 November 2007].

FAO (2011). Statistics Division 2013. available at: (http://faostat.fao.org/site/339/default.aspx) 20 July 2013, 15.30 PM.

Franks, J. (2013). Corn Syrup Greats: Delicious Corn Syrup Recipes, The Top 100 Corn Syrup Recipes. Emereo Publishing. p. 25. ISBN 978-1-4864-5955-1.

Gilani A. H., Mehmood M. H., Janbaz K. H., Khan A. U., Saeed S. A. (2008). Ethnopharmacological studies on antispasmodic and antiplatelet activities of *Ficus carica. J. Ethnopharmacol* : 119: 1-5.

Gordon, R. (2007). Lemon and Poppy Seed Scones with Homemade Lemon Curd. The Times (UK), June 20, 3:19pm.

Houda, L., Karima, S., Jean, C., Abdelwaheb, F. and Khaled, S. (2010). *In vitro* antimicrobial activity of four *Ficus carica* latex fractions against resistant human pathogens. *Pakistan Journal of Pharmaceutical Science*; 23(1): 53-58.

Howard, L. and Patten, M. (1960). The Australian Women's Weekly Cookery in colour, Paul Hamlin LTD, London UK, sections 956-971.

Isabel, D. W. and Schafer, W. (1990). Making Jams, Marmalades, Preserves, and Conserves. University of Minnesota extension school. Retrieved 20 September 2008.

Jeong, M. R, Cha, J. D., and Lee, Y. E. (2005). Antibacterial activity of Korean Fig (*Ficus carica* L.) against food poisoning bacteria. *Korean Journal Food Cookery Science.* 21:84-93

Khapre, A. P., Satwadhar, P. N and Deshpande, H. W. (2015). Studies on standardization of Fig fruit (*Ficus carica* L.) powder enriched cookies and its composition. *Asian Journal of Dairy and Food Research.* 34 (1) : 71-74.

Kirtikar, K. R. and Basu, B. D. (1996). Indian medicinal plants. International Book Distributors, India, Published by Biogreen Books, ISBN 13: 9788171360536

Mahalaskar, S. R., Lande, S. B., Satwadhar, P. N., Deshpande, H. W. and Babar, K. P. (2012). Development of technology for fortificatin of Fig (*Ficus carica* L.) fruit into its value added product- Fig toffee. *International Journal of Processing and Postharvest Technology*. 3 (2): 176-179.

McGovern, T. W. (2002). The Fig-*Ficus carica* L. Cutis; 69:339-40.

McMahon, M. J., Kofranek, A. M., and Rubatzky, V. E. (2002) Hartmann's Plant Science: Growth, Development, and Utilization of Cultivated Plants, 3rd ed.; Prentice Hall, Upper Saddle River, NJ, USA

Michailides, T. J. (2003). Diseases of Fig, p. 253–273. In: Ploetz, R.C. (ed.). Diseases of tropical fruit crops. CABI, Wallingford, UK.

Patten, M. (2001). Basic Basics: Jams, Preserves and Chutneys Handbook (2004 reprint ed.). Grub Street Books. ISBN 1-902304-72-1.

Rahemi, M. and H. Zare. (2002). The effects temperature treatments on disinfestation and storage of dry Figs of Estahban. *Journal of Science and Technology of Agriculture and Natural Resources*. 6: 29–41.

Rolek, B. (2015). Fruit Butters, Conserves, Curds, Jams, Jellies, Marmalades and Preserves. About.com website. http://easteuropeanfood.about.com/od/fruits/a/fruitbutters2.htm. Accessed May 11, 2015.

Solomon, A., Golubowicz, S., Yablowicz, Z., Grossman, S., Bergman, M., Gottlieb, H. E., Altman, A., Kerem, Z. and Flaishman M. A. (2006). Antioxidant activities and anthocyanin content of fresh fruits of common Fig (*Ficus carica* L.). *Journal of Agriculture and Food Chemistry*. 54 (20):7717-7723.

Stover, E., Aradhya, M. Ferguson, L. and Crisosto, C. H. (2007). The Fig: Overview of an Ancient Fruit. *Horticulture Science*. 42: 1083–1087.

Tanis, D. (2013). Fig and Almond Cake. The New York Times.

Tannahill, R. (1998). Food in History, Three Rivers Press, New York :49–51

Tanwar, B., Andallu, B. and Modgil, R. (2014). Influence of processing on physicochemical, nutritional and phytochemical composition of *Ficus carica* L. (Fig) products. *Asian Journal of Dairy and Food Research*. 33(1):37-43.

Turk, R., Eris, A., Ozer, M. H., Tuncelli, E. and Henze, J. (1994). Resaerch on the CA storage of Fig *cv*. Bursa Siyahi. *Acta Horticulture*. 368: 830-839.

Weigl, A. (2015). Learning to make a better Fig cake. The Charlotter Observer. September, 1st 2015: 8:31pm.

6

Jackfruit

Scientific Name: *Artocarpus heterophyllus* Lam.

Synonyms: *A. philippensis* Lamk., *A. maxima* Blanco

Family: Moraceae

Edible Portion: Bracts and Perianth

Other Common Names

Jackfruit, bo luop mi (China), jacquier (French), nanka (Indonesia), jaca, yaca (Spanish, Portuguese), lanka (Philippines), kapiak (New Guinea), uto ni India (Fiji), ulu initia (Samoa), chakka, chakki, kanthal, kathar, panos (India), jaca, jacca mole, jaca dura (Brazil), mit (Vietnam), khanun, makami, banum (Thailand)

Jackfruit (*Artocarpus heterophyllus*. L) is one of the underutilized fruits belonging to the family Moraceae and is the largest edible fruit in the plant kingdom (Morton, 1987), which occupies the top most rank with respect to quantity of food produced per unit area and available in plenty during the season (Haque, 2010). The total Jackfruit production in India has been recorded to be around 2.04 million tonnes (NHB, 2015). Out of this a significant portion goes waste beacuase of its highly perishable nature and seasonal glut. The postharvest losses in Jackfruit are around 30-35 per cent during the peak season (Lakshminarayan, 2017). Therefore, jack fruit has great potential for value addition for minimizing postharvest loses and enhancing the nonseasonal availablity. More than 100 items can be prepared from jack fruit right from immature stage to well ripened stage. The tree is valued for its money earning capacity and there are instances in which a single tree is reported to have generated an income of thousand rupees in one season alone. Hence, its cultivation is gaining popularity in the farming community. Owing to its multipurpose uses right from its roots to fruits, it is much-credited tree in the tropical world. Jackfruit is native to India and is mainly found in tropical Asian countries like India, Sri Lanka, Bangladesh, Indonesia, Malaysia and Philippines and it is quite popular in Eastern and Southern India and is cultivated widely in Kerala, Karnataka, Andhra Pradesh, Tamil Nadu, West Bengal, Maharashtra, Assam, Andaman and Nicobar Islands (Rahman *et al.,* 1999). The total area under Jackfruit in India is approximately 1.02 lakh hectares. It is cultivated in an area of about 0.11 lakh hectares mostly in Southern Plains and Western Ghats of Karnataka producing about 2.6 lakh tones of fruits per annum (Anon, 2000).

The true fruit is normally originated from carpel (ovary) and surrounded by a fleshy perianth, a main bulky portion of the fruit, which has three different regions *i.e.* middle fused region forms the rind of syncarp and the upper free horny region spikes. The lower fleshy edible portion is known as 'bulb'. It has very unique, pleasant aromatic flavor and distinct taste. The ripe fruits and seeds of Jackfruit are bestowed with certain chemical and aphrodisiac properties, which are important from health point of view. The nutritive value of Jackfruit is shown in Table 1 (Swamy *et al.,* 2012).

Importance

The fruit is rich in carotene, potassium and carbohydrates, moderately rich in ascorbic acid (Rahim and Quddus, 2000; Samaddar, 1985; Hossain *et al.,* 1979). It also contains some minerals like calcium, potassium and Vitamin B like thiamin, riboflavin, and Niacin (Acedo, 1992). The young and pre-mature fruit is also used as vegetable, which also contain high amount of vitamins and minerals. Seed is mainly used in curry, and reported to be more nutritious than the bulb, being richer in protein, fat, potassium and carbohydrate with considerable amount of phosphorus and calcium (Acedo, 1992; Rahim and Quddus, 2000). On the other hand, malnutrition problem is still prevailing in the country to a lesser extent particularly at rural and remote areas. The average food intake is deficient in calories, calcium, vitamin A, riboflavin and vitamin C. (INFS, 1989). It contains different bioactive compounds, such as carotenoids, flavonoids, total phenolics compounds, and ascorbic acid (Isabella *et al.,* 2010 and Jagadeesh *et al.,* 2007).

Jackfruit (*Artocarpus heterophyllus* lamk.) is known as the "**poor man's food**" (Molla *et al.,* 2008). It is generally used for culinary purposes in northern India as well as for table purpose in other parts of country. Ripe fruit can be processed for preparing the jam, squash, ready to serve beverages, chutney and papad. Fruits at early stage of the ripening may be utilized for making the preserve and candy. A puree prepared by adding water to the fruit in the ratio of 3:1 followed by passing through a pulper. The resulting puree was then mechanically pressed to obtain the juice (Seow and Shanmugam, 1992). Blended R.T.S. was prepared as per F.P.O. specifications (Giridhari lal *et al.,* 1995) by mixing fruit juice (100 ml) with the soya milk whey at 50:50 ratio, sugar and citric acid. The processed product was pasteurized at 80 ^{0}C (Saravan Kumar and Manimegalai, 2002). Squash was prepared by mixing syrup with 17 per cent juices to maintain the T.S.S of 52 ^{0}Brix and 1 per cent acidity (Bhatia *et al.,* 1995).

Nutritional and Medicinal Properties

Apart from its use as a table fruit, jack is a popular fruit for preparation of pickles, chips, leather and *papad.* The fruit has got good potential for value addition into several products like squash, jam, candy, halwa *etc.* The ripe bulbs can be preserved for one year in sugar syrup or in the form of sweetened pulp. The unripe mature bulbs can be blanched and dehydrated for further use through out the year. Seed is a rich source of starch and a delicacy during season. The timber is highly valued for its strength and sought for construction and furniture. The dried leaves are stitched to make disposable plates.

Jackfruit is rich in several nutrients and act as source of complete nutrition to the consumers (Table 1). The fruit is equivalent to Avocado and olive in terms of the healthier mix of nutrients for human dietary needs, almost having the exact nutrient equivalents of mother's milk. It is rich in vitamin B and C, potassium, calcium, iron, proteins and high level of carbohydrates, affordable and readily available supplement to our staple food. Its seeds are rich in proteins and can be relished as a nutritious nut. The fruit is also the source of major protein "Jacalin" from jack seed useful in preventing colon cancer and AIDS (Morton, 1987).

Health Benefits of Jackfruit

Strengthens Immune System

Jackfruit is an excellent source of vitamin C, which helps to protect against viral and bacterial infections. Vitamin C helps to strengthen the immune system function by supporting the white blood cells function. One cup of Jackfruit can supply the body with a very good amount of this powerful antioxidant.

Protects against Cancer

In addition to vitamin C, Jackfruit is also rich in phytonutrients such as *lignants, lisoflavones* and *saopnin, which* have anti cancer and anti aging properties. These phytonutrients may help to eliminate cancer causing free radicals from the body and slow down the degeneration of cells that can lead to degenerative diseases.

Aids in Healthy Digestion

Jackfruit is also known to contain anti ulcer properties, which help cure ulcers and digestive disorders. In addition, the presence of high fiber in the Jackfruit prevents constipation and helps in smooth bowel movements. These fibers also offer protection to mucous membrane by driving away the carcinogenic chemicals from the large intestine.

Maintains a Healthy Eye and Skin

Jackfruit contains vitamin A, a powerful nutrient that is known to maintain a healthy eye and skin. It also helps prevent vision related problems such as macular degeneration and night blindness.

Table 1: Composition of Jackfruit (100 g edible portion)

Sl.No.	*Composition*	*Young Fruit*	*Ripe Fruit*	*Seed*
A.	**Proximate Analysis**			
1	Water (g)	76.2-85.2	72.0-94.0	51.0-64.5
2	Protein (g)	2.0-2.6	1.2-1.9	6.60-7.04
3	Fat (g)	0.1-0.6	0.1-0.4	0.40-0.43
4	Carbohydrate (g)	9.4-11.5	16.0-25.4	25.8-38.4
5	Fiber (g)	2.6-3.6	1.0-1.5	1.0-1.5
6	Total sugar (g)	-	20.6	-
B	**Mineral and Vitamins**			
7	Total minerals (mg)	0.9	0.87-0.90	0.9-1.2
8	Calcium (mg)	30.0-73.2	20.0-37.0	50.0
9	Magnesium (mg)		27.0	54.0
10	Phosphorus (mg)	20.0-57.2	38.0-41.0	38.0-97.0
11	Potassium (mg)	287-323	191-407	246
12	Sodium (mg)	3.0-35	2.0-41.0	63.2
13	Iron (mg)	0.4-1.9	0.5-1.1	1.5
14	Vitamin A (IU)	30	175-540	10-17
15	Thiamine (mg)	0.05=0.15	0.03-0.09	0.25
16	Riboflavin (mg)	0.05-0.2	0.05-0.40	0.11-0.30
17	Vitamin C (mg)	1.0-14.0	7.0-10.0	11.0

Source: Swamy ***et al.,*** 2012.

Boosts Energy

Jackfruit is considered as an energy generating fruit due to the presence of simple sugars like fructose and sucrose, which give you an almost immediate energy boost. Although Jackfruit is an energy rich fruit, it contains no saturated fatty oil and cholesterol making it a healthy fruit to savor.

Lowers High Blood Pressure

Potassium contained in Jackfruit has been found to be helpful in the lowering of blood pressure and thus reducing the risk of heart attack as well as strokes.

Controls Asthma

The root of Jackfruit has been found to help those who suffer from asthma. Extract of boiled Jackfruit root has been found to control asthma.

Strengthens the Bone

Jackfruit is rich in magnesium, a nutrient, which is important in the absorption of calcium and works with calcium to help strengthen the bone and prevent bone related disorders such as osteoporosis.

Prevents Anemia

Jackfruit also contains iron, which helps to prevent anemia and also helps in proper blood circulation in our body.

Maintains a Healthy Thyroid

Jackfruit is loaded with copper, an important mineral, which plays a key role in the thyroid metabolism, especially in hormone production and absorption.

Table 2: Comparative Nutritional Status of some Fruits

Name of Fruit	*Ascorbic Acid (mg 100g^{-1})*	*Carotene (µg 100 g^{-1})*	*Sugar (per cent)*	*Brix (per cent)*	*Acidity (per cent)*
Mango	79.45	8.35	11.92	23.00	0.10
Jackfruit	7.04	175.0	16.46	21.50	0.25
Golden apple	87.12	1.60	6.33	9.00	1.40
Carambola	64.98	0.33	4.98	5.00	2.10
Custard apple	40.26	0.01	14.66	20.00	0.21
Indian olive	47.52	0.45	2.36	9.00	2.21
Aonla	425.0	2.39	6.87	13.00	2.35
Bael	10.24	55.33	16.83	24.50	0.32
Tamarind	43.48	0.03	8.03	10.50	10.32

Source: Bhuyan ***et al.,*** 2013.

Maturity

The maturity of Jackfruits is defined by the following parameters:

Coloring

Light green to yellow skin with light brown markings, yellow flesh. Cooking varieties are mature when skin is yellow with brown markings, fruit remains firm. Fresh eating varieties are mature when skin and stem is slightly yellowing; starch levels are low, flesh softening.

Shape

Generally oval to oblong, slightly irregular shape.

General Appearance

Rough skin composed of small spiny segments; skin often with brownish (latex stains; short stem present with clean cut.

Sensory

Starchy brown seeds surrounded by sweet aromatic flesh, which divides easily into segments; flesh texture may be soft and mucilaginous or firm depending on variety, when fully ripe the fruit emits a strong odour.

Size

<500 mm long and 250 mm wide

Classes

Jackfruit is classified in three classes defined below:

"Extra" Class

Jackfruit in this class must be of superior quality. It must be characteristic of the variety and/or commercial type. It must be practically free of defects provided that these do not affect the general appearance of the produce, the quality, the keeping quality and presentation in the package.

Class I

Jackfruit in this class must be of good quality. It must be characteristic of the variety and/or commercial type. The following defects, however, may be allowed, provided these do not affect the general appearance of the produce, the keeping quality and presentation in the package:

(a) Slight defects of colour and shape;

(b) Slight defect on the skin and other superficial defects not exceeding 10 per cent of the total surface area. The defects must not, in any case, affect the flesh of the fruit.

Class II

This class includes Jackfruits, which do not qualify for inclusion in the higher classes, but satisfy the minimum requirements. The following, however, may be allowed, provided the Jackfruit retains its essential characteristics as regards the quality, the keeping quality and presentation:

(a) Defects in shape and colour; (defects on the skin and other superficial

(b) Defects not exceeding 15 per cent of the total surface area.

Provisions Concerning Sizing

Size is determined by weight with the following table:

Size Code	*Weight (kg)*	*Size Code*	*Weight (kg)*
1	>25	4	10-15
2	21-25	5	<10
3	16-20		

Harvesting and Yield

Normally Jackfruit starts producing fruits from 7th to 8th year onwards. Grafted plants start to yield from 4th to 5th year itself. In Singapore jack, even seedlings start bearing from 3rd year. Fruit matures 6 to 8 months after flowering in March to June. In higher elevation harvest extends upto September and in plains certain genotypes bear an off-season crop during October to December. The optimum stage of harvest maturity of Jackfruit has been reported to 90 to 110 days after the appearance of the spike (Anonymous, 1975). At this stage, the ripened fruit has good eating quality in terms of aroma, texture, sweetness and taste—the total soluble solids (TSS) content has reached about 24 ^{0}Brix, total sugar about 11–15 per cent, while the total titratable acidity (TTA) is about 0.3 per cent. It is recommended that fruit should be harvested in the morning, as field heat is still low and tolerable to the produce (Punan *et al.,* 2000). Depending on rainfall, irrigation, and tree age, Jackfruit can produce from 20 to 250 fruits per year, sometimes up to 500 fruits on old-growth trees. Harvest indicators include a hollow sound when tapped, change of skin color, increased odor, and a flattening of its spines. In some region, a week or two before harvest the fruit stem (peduncle) is sliced to drain latex, which is said to speed ripening and improve flavor. This technique is becoming increasingly popular in India and other areas. Commercial yields average 100 to 150 kg/year/ tree or 20 to 100 fruits/year/tree. Orchards in Malaysia report a yearly average of 17, 000 kg/hectare (Love and Paull, 2011). The fruit weight varies from 10 to 30 kg. Being highly perishable, Jackfruit bulbs have a limited shelf life. Consequently, there is a need to utilize this crop for production of shelf stable products.

Storage

Jackfruits are soft and delicate, are more prone to damage and spoilage during handling and storage. Due to their high perishability, the postharvest management required is also high. It starts right from harvesting, field handling, transportation to pack house, pre- cooling and subsequent storage. Cold chains are essential component of horticultural postharvest infrastructure. It ensures maintenance of freshness of produce for extended period of storage. The cultivation requires linking operations more closely and systematically, modernizing marketing infrastructure and technologies, capacity building of the cultivators, and strengthening the policy for better marketing (Pandian and Soundarajan, 2014). Jackfruit is normally not stored in cold storage. A storage life of about 6 weeks is expected when the temperature is 11.1 to 12.8 ^{0}C and humidity between 85 to 90 per cent (Kripal, 1972). The initial quality and storage of maturity at harvest are important factors on which the storage life depends. Mathur *et al.* (1952) studied the changes in chemical composition of Jackfruit during storage at different temperature and found that storage at 11.1 to 12.8 ^{0}C and RH 85 to 90 per cent for 42 days resulted in decrease in sucrose content from 9.5 to 5.0 per cent while, reducing sugar content increased from 2.0 to 5.98 per cent. Reduction in level of ascorbic acid content from initial 8.2 per cent to 3.5 per cent was also recorded.

Value Added Products

Value addition is the process if changing or transforming a product from its original state to a more valuable state (Sharma *et al.,* 2014). Value addition to agricultural products is the process of increasing the economic value and consumer appeal of an agricultural commodity. Various value-adding technologies such as processing and preservation techniques, dehydration and drying technology, freezing technology, packing, labeling, *etc.* can be applied to agricultural produce to increase its value. The value added products from unripe, half ripe and ripe Jackfruit and seed are discussed here:

Stage of Fruit	*Value Added Products*
Unripe fruit	Pickle, Chips, Papad, Brined Jackfruit, RTC Jackfruit, Dehydrated Jackfruit, Culinary preparations, Cutlets, Biryani
Half ripe fruit	Candy, preserve
Fully ripe fruit	Jam, Leather, Rind Jelly, Squash, Nectar, Canned Bulbs, RTE (Ready-to-eat) Bulbs, RTS (Ready-to-serve) Drinks, Chutney, Toffee, Wine, Halwa, Kheer, GulabJamun, Icecream, Custard, Cake, Freeze dried pulp
Seeds	Seed powder, Starch flour, Culinary preparation, Pakoda, Kheer

Preservation of Jackfruit in Salt Solution

- ☆ Mature green Jackfruits are washed with clean water, peeled and cut into small pieces.
- ☆ Then kept them in 8 per cent salt, 1.25 per cent acetic acid, 0.1 per cent KMS and 91.65 per cent water solution (Bhuyan *et al.,* 2013).
- ☆ Then the materials poured into airtight plastic container and keep in cool and dry place.

Preservation of Jackfruit Bulb in Sugar Syrup

- ☆ Select fully ripen Jackfruit
- ☆ Wash with clean water
- ☆ Extract the bulbs and remove seeds from the bulb
- ☆ Make sugar solution with 25 per cent sugar and 0.5 per cent citric acid
- ☆ Add the bulbs into sugar solution
- ☆ Sterilize the bottles
- ☆ Pour the bulb into sugar solution containing bottle
- ☆ Store at room temperature (28-32 ^{0}C)

Osmotic Dehydrated Jackfruit Bulb

- ☆ The Jackfruits were washed, peeled, removed the seed from bulb and sliced lengthwise.

- Slices of the Jackfruits were divided into four groups and dipped into 45 ^{0}Brix sugar syrup concentrations and kept 30 minutes for soaking.
- Then, the slices were heated at 80 ^{0}C for 45 minutes.
- Finally, the slices were kept for 3 hrs and then soaked with 200 ppm potassium metabisulphite (KMS) according to FPO specification (Ranganna, 2007).
- The slices were transferred to drying trays and dehydrated at 50 ^{0}C for 24 hrs., 55 ^{0}C for 24 hrs and finally 60 ^{0}C for 8 hrs. The time and temperature of drying of the Jackfruit slices in the mechanical dryer was fixed up after several trials.
- After dehydration, the slices were packed in high-density polyethylene packet and stored at ambient temperature (Rahman *et al.,* 2012).

Ready to Cook (RTC) Tender Jack

Ingredients

- Jack fruit (tender/immature) - 1 kg
- Sodium hypochlorite - 25-50 g
- Sterilized water – 21 ml

Method

- Select good quality tender jack
- Remove outer peel using a clean knife
- Wash with sterile water with 10 per cent sodium hypochlorite
- Cut the treated fruit into small pieces
- Blanch the pieces for 3 minutes in hot water
- Pack the pieces in LDPE bags using hand wrapper
- It can be stored below 18 ^{0}C for 10 days

Source: Srivastava *et al.* (2017)

Tender Jack Bhaji

Sl.No.	*Ingredients*	*Quantity*	*Sl.No.*	*Ingredients*	*Quantity*
1	Jack fruit (tender/ immature)	250 g	6	Jaggery	10 g
2	Green chilies	5 to 6	7	Salt	As par taste
3	Mustard seeds	2 teaspoon	8	Oil	20 ml
4	Asafoetida	1 teaspoon	9	Coconut (Grated)	1 table spoon
5	Curry leaves	10-15 leaves			

Source: Devi *et al.,* 2014.

Method

- ✰ Heat oil in pan
- ✰ Season with mustard seeds. After they burst, add asafoetida, curry leaves, followed by green chillies and stir well till fried
- ✰ After the above get fried well, add finely chopped Jackfruit pieces, coconut, jaggery and salt to taste
- ✰ After the Jackfruit pieces get cooked well, it can be consumed.

Jackfruit Cutlet

Sl.No.	Ingredients	Quantity	Sl.No.	Ingredients	Quantity
1	Jack fruit (tender/ immature)	½ kg	10	Coriander leaves	½ bunch
2	Onion	3 No.	11	Ginger	10 g
3	Green chilli	6 No.	12	Garlic	6-7 clove
4	Pudina	5 g	13	Garam Masala	1 table spoon
5	Curry leaves	2-3 spring	14	Ginger garlic paste	2 tsp.
6	Turmeric powder	½ tsp.	15	Gram flour	3 tsp.
7	Red chilli powder	1 tsp.	16	Egg	2 No.
8	Oil	500 ml	17	Bread crumbs	Make crumbs by beating 6-7 slices of bread in mixer
9	Salt	As par taste			

Method

- ✰ Cut open mature unripe Jackfruit.
- ✰ Remove bulbs and extract out the seeds.
- ✰ Chop the bulbs into small pieces
- ✰ Chop onion, garlic, ginger, coriander leaves, pudina and curry leaves into small pieces
- ✰ Heat oil and fry onion followed by garlic, ginger, coriander leaves, pudina, curry leaves, green chillies and ginger-garlic paste. Then add Jackfruit pieces. Allow it to cook till soft. Then, allow it to cool for 10 minutes.
- ✰ Add garam masala, gram flour, red chilli powder, turmeric powder and salt to taste, to the above mixture.
- ✰ Take egg white and beat.
- ✰ Moisten your palm, make small balls of Jackfruit-cooked mixture, flatten the ball by gently pressing and dip in egg white.
- ✰ Roll it in bread crumbs.
- ✰ Deep fry in oil and serve hot.

Jackfruit Papad

Ingredients

- ✰ Mature/raw jack fruit bulbs - 500 g
- ✰ Salt - 2 tsp.
- ✰ Black sesame or cumin seeds - 2 tsp.

Method

- ✰ Cut open mature unripe Jackfruit. Remove bulbs and extract out the seeds.
- ✰ Boil the bulbs, drain well and grind into fine paste along with salt using mixer
- ✰ Mix ingredients like sesame or cumin seeds
- ✰ Flatten into layer of fine thickness
- ✰ Dry in trays of electric or solar cabinet drier/by open sun drying
- ✰ Deep fry and serve

Jackfruit Chips

Ingredients

- ✰ Well matured jack fruit bulbs - 1 kg
- ✰ Salt - to taste
- ✰ Water - to blanch
- ✰ Oil - 500 ml

Technical Flow-Chart for Processing of Chips

Select green mature Jackfruit with medium hard flesh

↓

Wash with clean water

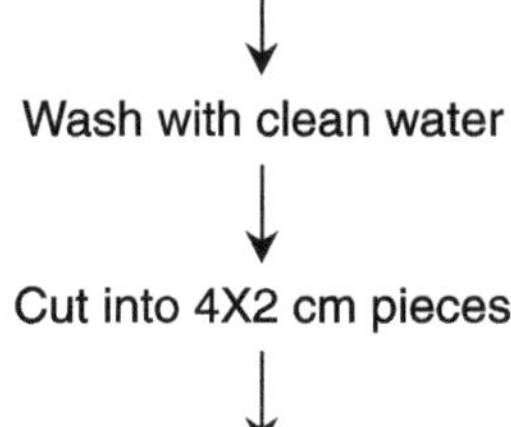

↓

Cut into 4X2 cm pieces

↓

Blanch the pieces into hot water at 95 °C for 5 min

↓

Dried the slices at 60 °C for 1 hrs and 70 °C for 6 hrs

↓

Finally the slices are fried at 160 °C in palm oil and stirred with narrow stick

↓

Put out the slices from pan when turn to light yellow colour and mix with tasting salt

↓

Pour the product into aluminium foil pouch

↓

Stored the product at room temperature (28-32 °C)

Source: Molla *et al.*, 2008.

Method

- Cut open mature unripe Jackfruit. Remove bulbs and extract out the seeds.
- Cut the bulbs into shreds of 0.5 to 0.6 cm width, maintaining the length as much as the bulb.
- Blanch the pieces in boiling water in which salt has been added for two minutes and allow to drain till completely dry.
- Heat oil in a frying pan and fry the chips. Add 1-2 spoons of salt water in oil while frying.
- After this, the processed products were packed in different packaging materials, namely i) Polypropylene pouch ii) High density polyethylene pouch (HDPE Pouch) and iii) Metalex foil pouch and stored in ambient temperature (28-32 °C) (Molla *et al.*, 2008).
- According to Molla *et al.* (2008) reported that moisture content (per cent), weight gain (per cent), quality aspects and sensory attributes like crispiness, colour, flavour and overall acceptability, metalex foil pouch was found most suitable for packaging of Jackfruit chips and prepared chips can be stored at ambient condition keeping in metalex foil for two months without loss of organoleptic quality.
- For vacuum frying of Jackfruit, the frying condition (pressure and temperature, frying rate and organoleptic test) were investigated by Alamsyah *et al.* (2002) and found that the Jackfruit are conducted under vacuum pressure of 70 cm/Hg and temperature level of 75 °C and 80 °C which, minimize the heat used and therefore reduce changes in composition, color, taste and flavor of the Jackfruit.

Jackfruit Bajjas

Sl.No.	*Ingredients*	*Quantity*	*Sl.No.*	*Ingredients*	*Quantity*
1	Jack fruit (tender/ immature)	500 g	5	Pepper powder	1 tsp.
2	Gram flour	250 g	6	Cumin powder	½ tsp.
3	Chilli powder	2 tsp.	7	Salt	As par taste
4	Turmeric powder	1 tsp.	8	Oil	500 ml

Source: Devi *et al.*, 2014.

Method

- ☆ Cut open well mature unripe Jackfruit. Remove bulbs and extract out the seeds. Cut one bulb into two pieces
- ☆ Add salt, chilli powder, turmeric powder, jeera powder and pepper powder to gram flour and mix well by adding water till medium consistency.
- ☆ Dip the fruit pieces into the batter
- ☆ Heat oil in frying pan and deep fry in hot oil till they turn golden brown.

Jackfruit Pakodas

Sl.No.	*Ingredients*	*Quantity*	*Sl.No.*	*Ingredients*	*Quantity*
1	Jack fruit (tender/ immature)	500 g	6	Gram flour	200 g
2	Onion	3 No.	7	Salt	As par taste
3	Green chilli	8 No.	8	Chilli powder	2 tsp.
4	Coriander leaves	Half bunch	9	Oil	500 ml
5	Curry leaves	2-3 sprigs			

Source: Devi *et al.*, 2014.

Method

- ☆ Cut open well mature unripe Jackfruit. Remove bulbs and extract out the seeds.
- ☆ Shred the bulbs into small pieces, add chopped onion, green chilli, curry leaves and coriander leaves.
- ☆ Add gram flour, salt and chilli powder to the above; mix well by adding water till required consistency.
- ☆ Make small to medium sized balls and deep fry in oil

Jackfruit Xacuti

Sl.No.	*Ingredients*	*Quantity*	*Sl.No.*	*Ingredients*	**Quantity**
1	Jack fruit (tender/ immature)	1 kg	4	Chilli powder	20 g
2	Oil	250 ml	5	Tomato	150 g
3	Onion	200 g	6	Turmeric powder	5 g

Source: Devi *et al.*, 2014.

Method to Prepare Garam Masala

- ☆ Dry roast all ingredients required for garam masala and powder them for later use.

Preparation of Wet Masala

Ingredients

- ☆ Coconut (grated) - 1 full (medium size)
- ☆ Chopped onion - 100 g
- ☆ Garlic - 10 pods
- ☆ Poppy seeds - 1 tsp.
- ☆ Tamarind - Small lemon size

Method to Prepare Wet Masala

- ☆ In a pan, heat oil, add 100 g of onion, garlic pods and fry till golden brown. To it, add grated coconut and fry till light brown followed by poppy seeds and keep aside.
- ☆ Grind fried coconut, onion and garlic with tamarind into smooth paste using water

Method to Prepare Xacuti

- ☆ Peel Jackfruit and make into big pieces (like chicken pieces) and blanch in water for 5 mins
- ☆ Heat oil and fry jack fruit pieces till golden brown and keep aside.
- ☆ In a vessel, heat oil, fry 100 g of finely chopped onion till brown. Then, add 100 g of chopped tomato and fry well.
- ☆ To this, add garam masala powder, chilli powder and then add fried jack pieces,
- ☆ Mix well and then add the coconut based wet masala and turmeric powder. Stir well and cook for 10 mins.
- ☆ Garnish with chopped coriander leaves and serve with hot rice or roti.

Jackfruit Pulp (Basic recipe)

Sl.No.	Ingredients	Quantity	Sl.No.	Ingredients	Quantity
1	Jack fruit (pulp) (tender/immature)	1 kg	3	Ghee	200 g
2	Jaggery	1 kg	4	Water	1 ltr.

Method

- ☆ Cut open well ripe Jackfruit. Remove bulbs and extract out the seeds.
- ☆ Boil the ripe bulbs with minimum required water till they turn soft.
- ☆ Drain the excess water and grind the bulbs into a smooth pulp using mixer.
- ☆ Prepare jaggery syrup by dissolving one kg of jaggery in one litre of water.

- ☆ After the syrup boils add Jackfruit pulp and ghee.
- ☆ Mix well and stir continuously till the pulp is thick in consistency.

Jackfruit Halwa

Sl.No.	*Ingredients*	*Quantity*	*Sl.No.*	*Ingredients*	*Quantity*
1	Jack fruit (pulp)	200 g	4	Cashewnuts	10 No.
2	Water	1 cup	5	Maida	1 tsp.
3	Ghee	½ cup	6	Sugar	1 cup

Method

- ☆ In a broad vessel with thick base, add sugar, water and maida and mix well. When it starts boiling, add cardamom, ghee roasted cashewnut and ghee.
- ☆ Stir till it thickens to consistency of halwa.
- ☆ Apply ghee to tray/plate and spread.
- ☆ Allow it to cool, cut and serve.

Gulab Jamum

Sl.No.	*Ingredients*	*Quantity*	*Sl.No.*	*Ingredients*	*Quantity*
1	Jack fruit (pulp)	5 tsp.	4	Cardamom powder	1 tsp.
2	Milk powder	2 cup	5	Maida	2 tsp.
3	Ghee	1 tsp.	6	Sugar	250 g

Method

- ☆ Add maida and milk powder to jack fruit pulp.
- ☆ Add 2-3 drops of vanilla essence and cardamom powder and make smooth dough.
- ☆ Make sugar syrup by adding 250 g of sugar in 250 ml of water. After the syrup boils, simmer the flame and leave for 5 mins to get the required thin consistency sugar syrup.
- ☆ Make small balls of dough, deep fry in ghee and soak in sugar syrup

Jackfruit Unniyappam (Mini Appams)

Sl.No.	*Ingredients*	*Quantity*	*Sl.No.*	*Ingredients*	*Quantity*
1	Jack fruit (pulp)	250 g	4	Cardamom powder	2 tsp.
2	Black sesame	50 g	5	Coconut	½ (finely chopped into cubes)
3	Rice flour	250 g	6	Salt	½ to 1 tsp.

Source: Devi *et al.*, 2014.

Method

- ☆ Fry the finely chopped coconut cubes/pieces in ghee.
- ☆ Add rice flour, sesame seeds, cardamom powder, roasted coconut pieces and pinch of salt to basic recipe.
- ☆ Mix well the above ingredients into a smooth dough, without adding water.
- ☆ Roll into small balls and deep fry in oil.

Jackfruit Sweet Vada

Sl.No.	*Ingredients*	*Quantity*	*Sl.No.*	*Ingredients*	*Quantity*
1	Jack fruit (pulp)	250 g	4	Sesame seeds	50 g
2	Maida flour	250 g	5	Cardamom powder	2 tsp.
3	Baking powder	1 tsp.	6	Salt	½ to 1 tsp.

Source: Devi *et al.*, 2014.

Method

- ☆ Add maida flour, sesame seeds, cardamom powder, baking powder and pinch of salt to basic recipe.
- ☆ Mix all the above ingredients into a smooth dough, without adding water.
- ☆ Roll into balls, flatten on a greased surface, pierce a hole with finger and deep fry in oil.

Jackfruit Custard

Sl.No.	*Ingredients*	*Quantity*	*Sl.No.*	*Ingredients*	*Quantity*
1	Jack fruit (pulp)	500 g	4	Sugar	500 g
2	Custard powder	4 tsp.	5	Milk	½ ltr.

Source: Devi *et al.*, 2014.

Method

- ☆ Add custard powder to milk, mix well and boil it.
- ☆ Add sugar to boiling milk and boil for few more minutes.
- ☆ After the milk cools to room temperature, refrigerate it for 4-5 hours/till the required semi solid consistency is reached.
- ☆ Cut ripe jack fruits into small pieces, add to the custard and serve.

Jackfruit Wine

Sl.No.	*Ingredients*	*Quantity*	*Sl.No.*	*Ingredients*	*Quantity*
1	Ripe jack fruit (pulp)	1 kg	6	Poppy seeds	10 No.
2	Sugar	500 g	7	Cardamom	2-3 No.
3	Water	1 ltr	8	Star anise	1 No.
4	Yeast	½ tsp.	9	Cloves	2-3 No.
5	Cinnamon	½ tsp.			

Technical Flow-Chart for Processing of Wine

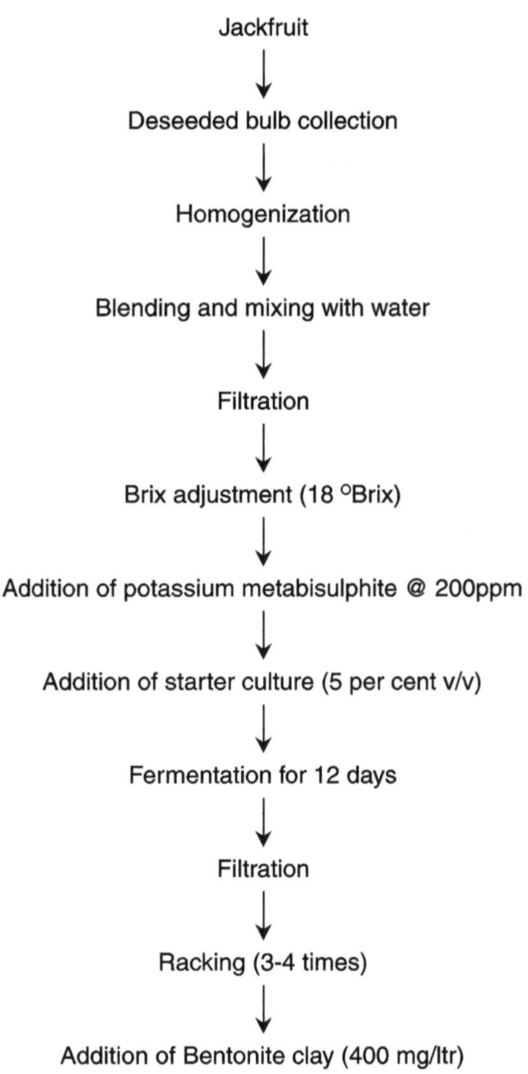

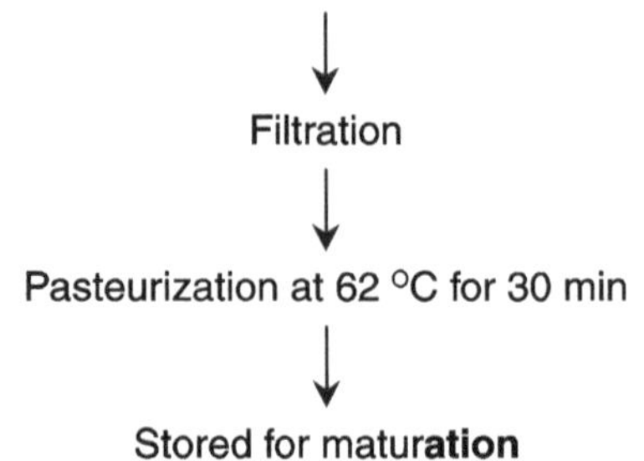

Source: Dushyantha *et al.*, 2011; Kumoro *et al.*, 2012.

Method

- The extracted juice was ameliorated to obtain 18 °Brix by adding 200 g cane sugar per liter of juice.
- The acidity was maintained at 0.5 per cent by adding citric acid.
- Wild yeasts present in the juice were suppressed by adding Potassium metabisulphite @ 200 ppm.
- The juice was inoculated with lactic acid bacterial starter cultures @ 5 per cent (v/v) for fermentation.
- Fermentation was carried out with occasional mixing of the juice and the fermentation was stopped by fall in T.S.S (°Brix).
- The wine was filtered using cheese cloth and filled in bottles for completion of slow fermentation. Finally, wines were clarified by adding 0.4 per cent Bentonite clay.
- Further, clear wine was siphoned into clean pasteurized bottles and tightly corked.
- According to Kumoro *et al.*, 2012 fermentation of original Jackfruit juice of 14 per cent w/w sugar concentration using 0.5 per cent w/v yeast for 9 days was the best to produce a good quality wine with 12.13 per cent v/v of ethanol and specific Jackfruit aroma.

Jackfruit Kheer

Sl.No.	*Ingredients*	*Quantity*	*Sl.No.*	*Ingredients*	*Quantity*
1	Ripe jack fruit (pulp)	500 g	5	Cardamom powder	1 tsp.
2	Rice flour	50-100 g	6	Jaggery	200 g
3	Coconut	1 No.	7	Dry Fruits	100 g
4	Milk	250 ml	8	Ghee	100

Method of Extracting Coconut Milk

- Grate the coconut.

- ☆ Put the grated coconut into a mixer jar, add equal volume of warm water and run the mixer for 15 seconds.
- ☆ It can be done in batches if the mixer jar is too small for the grated coconut.
- ☆ Extract the coconut milk by squeezing the ground coconut. This is called the first coconut milk. Keep it aside. Pool all the first milk extracted if done in batches.
- ☆ Then, add the coconut residue that is left out after the extraction of the first coconut milk into the mixer jar. Add equal volume of warm water and run the mixer for 10-15 seconds.
- ☆ Again repeat the extraction by squeezing the ground coconut residue. This is called the second coconut milk. Pool all the second milk extracted if done in batches.

Jackfruit Pudding

Sl.No.	*Ingredients*	*Quantity*	*Sl.No.*	*Ingredients*	*Quantity*
1	Ripe jack fruit (pulp)	500 g	4	Milk	500 ml
2	Sugar	250 g	5	China grass	20 g
3	Cream/condensed milk	150 g	6	Vanilla essence	1 tsp.

Source: Devi *et al.*, 2014.

Method

- ☆ Cut china grass into small pieces. Add in a glass of water and keep it aside.
- ☆ Boil milk, sugar and the china grass water.
- ☆ Cook and pulp the well ripe Jackfruit.
- ☆ Add the pulp to the milk mixture and stir on a slow flame till a thick custard consistency is formed. Add cream and essence.
- ☆ Cool in the fridge for about 2 hrs and serve.

Jackfruit Fritters

Sl.No.	*Ingredients*	*Quantity*	*Sl.No.*	*Ingredients*	*Quantity*
1	Ripe jack fruit (pulp)	200 g	5	Sesame seeds	1 tsp.
2	Maida	50 g	6	Cardamom powder	¼ tsp.
3	Corn flour	25 g	7	Oil	200 ml.
4	Sugar	1 tsp.	8	Salt	As par taste

Source: Devi *et al.*, 2014.

Method

- ✰ Cut ripe Jackfruit bulbs into broad pieces.
- ✰ Mix all the ingredients except oil into a thick batter.
- ✰ Dip the Jackfruit pieces in the batter.
- ✰ Fry in the oil till golden brown and serve

Jackfruit Squash

Sl.No.	Ingredients	Quantity	Sl.No.	Ingredients	Quantity
1	Ripe jack fruit (pulp)	1 kg	4	Water	100-150 ml
2	Pineapple	1 No.	5	Citric acid	1 tsp.
3	Sugar	¾ cup	6	Vanilla essence	1 tsp.

Method

- ✰ Cut open well ripe Jackfruit. Remove bulbs and extract out the seeds.
- ✰ Cut the well ripe bulbs into small cubes.
- ✰ Boil 1 kg of pieces in water and then pulp into fine paste.
- ✰ Take 1 part of pulp, add water and mix thoroughly using mixer.
- ✰ Extract juice from pineapple and add to jack pulp in 0.5: 1 ratio.
- ✰ Prepare sugar syrup by boiling 250 g of sugar in 250 ml of water.
- ✰ Add sugar syrup to jack fruit pulp followed by citric acid.
- ✰ Add 700 mg of Potassium Meta bisulphite for 1 litre of squash.
- ✰ Cool and fill in glass bottles

Jackfruit Jam

Sl.No.	Ingredients	Quantity	Sl.No.	Ingredients	Quantity
1	Ripe jack fruit (pulp)	1 kg	4	Citric acid	8 g
2	Sugar	750 g	5	Pectin	1 g

Source: Mondal ***et al.*** (2013)

Technical Flow-Chart for Processing of Jam

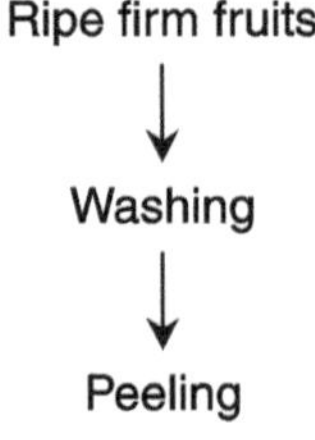

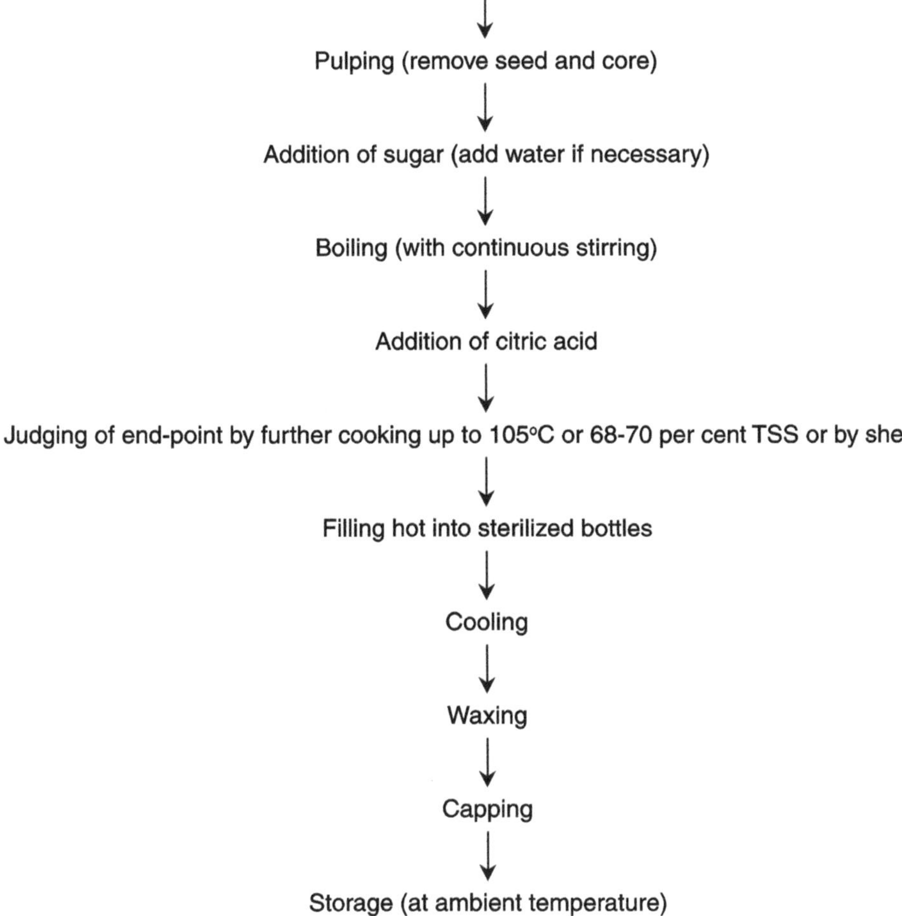

Method

- Cut the well ripe jack fruit into small pieces.
- Boil the pieces with water and pulp into fine paste.
- Add the jack fruit paste and sugar and cook on pan with little water, if required
- Add permitted food colour and citric acid to the mixture. Add one tablespoon of lime juice to the same (optional).
- Stir continuously till jam consistency.
- Test for end point using ladle test/flake test/sheet test.

Jackfruit Leather

Ingredients

- ☆ Well ripened jack fruit - 500 g

Method

- ☆ Cut the well ripened jack fruit into small pieces
- ☆ Pulp into fine paste
- ☆ Spread the smooth pulp as uniform layer on trays
- ☆ Dry using solar or electric cabinet drier. The leather can also be dried under direct sun light in plates or trays. It is dried till moisture is lost and starts coming out of the tray. If dried beyond this, it becomes brittle.
- ☆ After drying, cut into desired size and shape and pack in polythene pouches.
- ☆ Okilya *et al.* (2010) reported that the solar drying reduced moisture content to desired levels and had similar effects on texture when compared to cabinet and oven drying, solar dried leather was generally not sensorially acceptable and not acceptable because its long drying time and low temperature encouraged greater loss of colour pigments and aroma compounds as well as browning reactions.

Jackfruit Chocolate

Sl.No.	*Ingredients*	*Quantity*	*Sl.No.*	*Ingredients*	*Quantity*
1	Ripe jack fruit (pulp)	1 kg	4	Butter	100 g
2	Sugar	500 g	5	Cocoa powder	50 g
3	Milk powder	150 g			

Source: Devi *et al.*, 2014.

Method

- ☆ Cut the well ripened jack fruit into small pieces.
- ☆ Pulp into fine paste.
- ☆ Mix jack fruit pulp and sugar in a heavy bottom pan.
- ☆ Cook on flame till it reduces to $1/3^{rd}$ volume
- ☆ Then add milk powder dissolved in 100 ml of hot water, followed by ghee and cocoa powder mixed in hot water.
- ☆ Mix well and stir on low fire till desirable consistency. One should be able to make balls by rolling on palm.
- ☆ Pour the mixture in a plate and roll it into chocolates.
- ☆ Refrigerate for 2 hrs and then pack individually in butter paper.

Jack Seed Payasam/Kheer

Sl.No.	*Ingredients*	*Quantity*	*Sl.No.*	*Ingredients*	*Quantity*
1	Jack seed	500 g	4	Cardamom	3 No.
2	Jaggery	300 g	5	Ghee	100 g
3	Coconut	1 No.			

Method of Extracting Coconut Milk

- ✰ Grate the coconut.
- ✰ Put the grated coconut into a mixer jar, add equal volume of warm water and run the mixer for 15 seconds.
- ✰ It can be done in batches if the mixer jar is too small for the grated coconut.
- ✰ Extract the coconut milk by squeezing the ground coconut. This is called the first coconut milk. Keep it aside. Pool all the first milk extracted if done in batches.
- ✰ Then, add the coconut residue that is left out after the extraction of the first coconut milk into the mixer jar. Add equal volume of warm water and run the mixer for 10-15 seconds.
- ✰ Again repeat the extraction by squeezing the ground coconut residue. This is called the second coconut milk. Pool all the second milk extracted if done in batches.

Jack Seed Burfi

Sl.No.	*Ingredients*	*Quantity*	*Sl.No.*	*Ingredients*	*Quantity*
1	Jack seed	25-30 No.	6	Butter	150 g
2	Maida	100 g	7	Vanilla essence	1 tsp.
3	Gram flour	150 g	8	Cardamom powder	¼ tsp.
4	Milk	200 ml	9	Pista crushed	1 tsp.
5	Sugar	500 g			

Method

- ✰ Cut open well ripe Jackfruit. Remove bulbs and extract out the seeds.
- ✰ Boil the jack fruit seeds till they are cooked.
- ✰ Remove seed coat manually, cut into small pieces and make a smooth paste using a mixer.
- ✰ Add maida, gram flour and sugar to milk and mix well.
- ✰ Mixe thoroughly Jackfruit seed paste, milk mixture into a pan.
- ✰ Keep on the fire and keep on stirring till it becomes hard and roll into a ball in the hand.

- Keep adding butter in small quantities while stirring.
- Towards the end add vanilla essence, cardamom powder and pista.
- When done pour the mixture into a greased pan.
- When cool cut into pieces.

Jack Seed Pakodas

Sl.No.	Ingredients	Quantity	Sl.No.	Ingredients	Quantity
1	Jack seed	500 g	6	Gram flour	50 g
2	Onion	3 No.	7	Chilli powder	2 tsp.
3	Green chilli	8 No.	8	Oil	500 ml
4	Coriander leaves	Half bunch	9	Salt	As par taste
5	Curry leaves	2-3 springs			

Method

- Boil the jack fruit seed.
- Remove seed coat, cut into small pieces and make powder in a mixer.
- Add chopped onion, green chilli, curry leaves and coriander leaves. Add gram flour, salt and chilli powder to the above; mix well by adding water till required consistency.
- Make small to medium sized balls and deep fry in oil.

Preparation of Jackfruit Rind Flour (JRF)

- Jackfruit rind needed to be processed before drying and milling in order to obtain JRF.
- They were rinsed with running tap water, followed by deionized water and soaked in boiling water for 10 min to soften the texture and then rinsed.
- Thereafter, they were soaked in a boiling solution of 0.1 per cent w/w sodium bisulfite ($NaHSO_3$) for 10 min and rinsed.
- Pieces of Jackfruit rind were soaked in another boiling solution of sodium bicarbonate ($NaHCO_3$) for 15 min and rinsed three times.
- They were dried in convection dryer for 24 h at 50 ^{0}C. The dried samples were ground using a miller and further sieved through a 355-µm-mesh sieve.
- The obtained Jackfruit rind flour (JRF) were kept in airtight plastic containers and stored in a fridge at 4 ^{0}C priors to use (Feili *et al.,* 2013).

Jackfruit Powder

Method

- The Jackfruit was cut in half-length wise and the core of the fruit was carved out.
- The pulp was scooped and cut into 2 – 4 pieces while the seeds removed.
- The pulp was blanched for two minutes and the cooled rapidly under tap water.
- It was dried in the cabinet drier at 55 °C for 6 to 7 hour until the moisture content was reduced to 5 per cent.
- The dried pulp was ground into powder and sieved to remove lumps and unground materials.
- The Jackfruit powder was packed into moisture-proof container and heat sealed (Gunasena *et al.,* 1996; Kumar *et al.,* 2012).

Jackfruit Seed Flour

- The Jackfruit seeds were cleaned manually and white arils (seed coat) were manually peeled off.
- Seeds were lye peeled, soaking in 3 per cent sodium hydroxide solution for 3-5 minutes to remove the thin brown by rubbing the seeds between the hands and washing thoroughly under running water.
- Lye peeled seeds were sliced into thin chips and dried at 50- 60 °C 24 hrs. (Ocloo *et al.,* 2010).
- The dried chips were powdered in a flourmill, passed through 60 mm mesh sieve and packed in polyethylene pouches and stored in a refrigerator (<10 °C) temperature (Butool and Butool, 2015; Chowdhury *et al.,* 2012).
- Islam *et al.* (2015) stated that the Jackfruit seed flour can be utilized for making highly nutritious as well as low fat content biscuits and it was so better as compare to plain biscuits.

Extraction of Pulp

Fresh ripe fruit cuts into in four sections from vertical to horizontal section. Separate the cubes and remove the seeds and collect the fresh flesh. Pulping by pulper machine and collect the pulp. Filled in sterilized bottles with adding of preservative, capping and sealing. Pasteurized at 80 °C for 20 minutes. Stored at refrigeration or room temperature. In room temperature can be stored up to 10 month, but in refrigeration can be stored up to 2 years.

Jackfruit RTS (Ready to Serve)

Sl.No.	Ingredients	Quantity	Sl.No.	Ingredients	Quantity
1	Jackfruit juice	110 ml	4	Water	850 ml
2	Sugar	110 g	5	KMS	0.2 g/ltr of finished product
3	Citric acid	3 g			

Procedure

Preparation of Syrup

Boil the sugar with water in stainless steel utensil, and add the citric acid to the boil syrup for clarification. Strain the syrup and cool the syrup.

Add the syrup little by little to the mixed with pulp and stirring, and get final RTS. Filled in sterilized bottles with adding of preservative, capping and sealing. Pasteurized at 80 ^{0}C for 20 minutes. Stored at refrigeration or room temperature.

Jackfruit Pickle

The preservation of fruit or vegetables in common salt or vinegar is called pickling. Spices and oil may also be added in pickle. Common salt @ 15 per cent prevents its spoilage. Vinegar too @ 2 per cent acts as a preservative.

Sl.No.	Ingredients	Quantity	Sl.No.	Ingredients	Quantity
1	Jackfruit pieces	2 kg	10	Large cardamom	4 pieces
2	Salt	350 gram	11	Small cardamom	10 pieces
3	Turmeric powder	15 gram	12	Black pepper powder	10 gram
4	Nigela seed	40 gram	13	Coriander powder	10 gram
5	Red chilli powder	15 gram	14	Asafetida	5 gram
6	Fenugreek seed	60 gram	15	Onion	50 gram
7	Clove	10 pieces	16	Ginger	15 gram
8	Cumin powder	10 gram	17	Garlic	15 gram
9	Cinnamon powder	10 gram	18	Oils	800 ml

Source: Patil *et al.*, 2011; Tilak, 1995.

Technical Flow-Chart for Preparation of Pickle

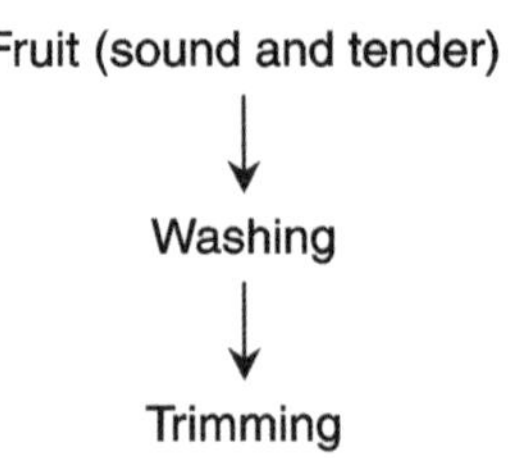

Peeling (rough and thick skin)
Cutting rectangle slices (1-1.5cm thick)
Blanching for 5 minutes
Draining off water
Drying in shade for an hour
Addition of salt and turmeric powder
Keeping in sun light for a week
Frying spices in a little oil (except pulp)
Mixing pieces with fried spices
Cooling
Add jaggery or vinegar (as per need)
Filling in jar
Addition of oil (heating and cooled it)
Storage

Jackfruit Candy

Sl.No.	Ingredients	Quantity	Sl.No.	Ingredients	Quantity
1	Jackfruit pulp	500 g	4	Potassium metabisulphite	2 g
2	Water	1250 ml	5	Citric Acid	3 g
3	Sugar	750 g			

Source: ICUC (2005).

Method

- The Jackfruit was cut into half lengthwise.
- The core of the fruit was carved out and the pulp was scooped out.
- The seeds were removed from the pulp.
- The pulp was soaked in a brine solution containing 15 per cent sodium chloride (150 g/liter) and 1 per cent calcium chloride (10 g/liter) for 2 days.
- The pulp was removed from the brine, drained and then washed with water to remove the salt.
- It was boiled at 14 ^{0}Brix sugar syrup for 5 minutes and then kept at room temperature (28.3 ^{0}C) for 24 hours. The pulp was removed from the syrup and then dried for about one day (24 hours) in a tray dryer.
- The Jackfruit candy was packed in cardboard cartons.
- According to Rahman *et al.* (2012) studied the preservation of Jackfruit by osmotic dehydration with different sugar concentration *viz.* 35, 40, 45 ^{0}Brix and found that the product prepared the solution of 45 ^{0}Brix was stored maximum time with highest organoleptic value.

REFERENCES

Acedo Jr., A. L. (1992). Jackfruit Biology, Production, Use, and Philippine Research. Monograph Number 1. Forestry/Fuelwood Research and Development (F/FRED) Project, Arlington, Virginia.

Alamsyah, R., Budhiono, A., Setiwan, Y. Y. and Rinaldi, H. (2002). Vacuum frying for Jackfruit: processing, financial and analysis. Warta IPH/Journal of Agro-Based Industry. 19 (1-2): 36-44.

Anonymous (1975). Annual Report, CFTRI, Mysore, India: 24-25.

Anonymous. (2000). Jackfruit processing-fruitful SSI venture, Indian Food Industry. 19: 323.

Bhatiya, B. S., Siddapa, G. and Lal, G. (1995). Development of products from Jackfruit: part III- Jackfruit preserve, candy, chutney, and dried bulb. *Indian Food Packer.* 9: 7.

Bhuyan, M. A. J., Saha, M. G. and Rahman, M. A. (2013). Value Addition to Jackfruit (*Artocarpus heterophyllus* Lam.) through Integrated Processing and Preservation. Workshop on Valorisation of traditional processing of indigenous and underutilized fruits, at Institute of Technology of Cambodia, Phnom Penh, Cambodia 14- 16th January, 2013.

Butool, S. and Butool, M. (2015). Nutritional quality on value addition to Jackfruit seed flour. *International Journal of Science and Research.* 4(4): 2406-2411.

Chowdhury, A. R., Bhattacharyya, A. K. and Chatopadhyay, P. (2012). Study on functional properties of raw and blended Jackfruit seed flour (a non conventional source) for food application. *Indian Journal of Natural Products and Resources.* 3 (3): 347-353.

Devi, S. P., Talaulikar, S., Gupta, M. J., Thangam, M. and Singh, N. P. (2014): A Guide on Jack Fruit - Cultivation and Value Addition. Technical Bulletin No. 41, ICAR (RC), Goa.

Dushyantha, D. K., Raghwendra, K. M. and Suvarna, V. C. (2011). Study on Fermentation of Jackfruit (*Artocarpus heterophyllus* L.) Juice by beneficial Lactic Acid Bacteria. *Research Journal of Pharmaceutical, Biological and Chemical Sciences.* 2 (4): 732-739.

Feili, R., Zzaman, W., Abdullah, W. N. W. and Yang, T. A. (2013). Physical and Sensory Analysis of High Fiber Bread Incorporated with Jackfruit Rind Flour. *Food Science and Technology.* 1 (2): 30-36.

Girdhari Lal, Siddapa, G. S. and Tandon, G. L. (1995). Preserves candied and crystallized fruits. In: *Presevation of Fruit and Vegetables.* ICAR, New Delhi.

Gunasena, H. P. M., Ariyadasa, K. P., Wikramasinghe, A., Heralin, H. M. W., Wikramasinghe, P. and Rajakarunas, B. (1996). *Manual of Jackfruit Cultivation and Utilization.* In Srilanka Forest Information Service, Forest Department, Sri Lanka.

Haque, M. A. (2010). Jackfruit: Botany, Production Technology and Research in Bangladesh. Jahanara Haque Publisher, 291, Dhaka Road, Mymen Singh: 143.

Hossain, M. M., Haque, A. and Hossain, M. (1979). Nutritive value of Jackfruit. *Bangladesh Journal of Agricultural Science.* 1 (2): 9-12.

ICUC (2005). Jackfruit Recipe. *Training Manual on Processing and Small Business Development,* International Centre for Underutilized Crops University of Southampton, U.K.

INFS (1989). Nutrition survey of rural Bangladesh. Institute of Nutrition and Food Science, Univ. Dhaka, Bangladesh.

Isabella, M., Lee, B. L., Lim, L. T., Koh, W. P. Huang, D. and Ong, C. N. (2010). Antioxidant activity and profiles of common fruits in Singapore". *Food Chemistry.* 120: 993-1003.

Islam, M. S., Begum, R., Khatun, M. and Dey, K. C. (2015). A study on nutritional and functional properties analysis of Jackfruit flour and value addition to biscuits. *International Journal of Engineering Research and Technology*. 4 (12): 139-147.

Jagadeesh, S. L., Reddy, B. S., Swamy, G. S. K., Gorbal, K., Hegde, L. and Raghavan, G. S. V. (2007). Chemical composition of Jackfruit (*Artocarpus heterophyllus* Lam.) selections of Western Ghats of India. *Food Chemistry*. 102 (1): 361-365.

Kripal, Singh. K. (1972). Farm Information Bulletin No. 71, Directorate of Extension, Ministry of Agriculture, New Delhi.

Kumar, V., Suneetha, P. K. and Sucharitha, K. V. (2012). Freez drying- A novel processing for fruit powders. *International Journal of Food and Nutritional Sciences*. 1(1): 16-29.

Kumoro, A. C., Sari, D. R., Pinandita, A. P. P., Retnowati, D. S. and Budiyati, C. S. (2012). Preparation of Wine from Jackfruit (*Artocarpus heterophyllus* Lam.) Juice Using Baker yeast: Effect of Yeast and Initial Sugar Concentrations. *World Applied Sciences Journal*. 16(9): 1262-1268.

Lakshminarayan, N. G. (2017). Jackfruit based value added products. Available at https://www.linkedin.com/pulse/Jackfruit-based-value-added-products-n-g-lakshminarayan, accessed on 02/05/2017.

Love, K. and Paull, R. E. (2011). Jackfruit. *Fruit and Nuts*. 19: 1-7.

Mathur, P. B., Singh, K. K. and Kapur, M. S. (1952). Indian Society of Refrigerated. Enginners. 3: 40-46.

Molla, M. M., Nasrin, T. A. A., Islam, M. N. and Bhuyan, A. J. (2008). Preparation and packaging of Jackfruit chips. *International Journal of Sustainable Crop Production*. 3 (6):41-47.

Mondal, C., Remme, R. N., Mamun, A. A., Sultana, S. and Ali, M. H. (2013). Product Development from Jackfruit (*Artocarpus heterophyllus*) and Analysis of Nutritional Quality of the Processed Products. *IOSR Journal of Agriculture and Veterinary Science*, 4(1): 76-84.

Morton, J. (1987). Jackfruit. *In*: Fruits of warm climates.: 58–64.

NHB (National Horticulture Board) (2015). Horticultural Statistics at a Glance 2015. Available at http://nhb.gov.in/ PDFViwer.aspx?enc=3ZOO8K5CzcdC/ Yq6HcdIxC0U1k ZZenFuNVXacDLxz28= as accessed on 02/05/2017.

Ocloo, F. C. K., Bansa, D., Boatin, R., Adom, T. and Agbemavor, W. S. (2010). Physico-chemical, functional and pasting characteristics of flour produced from Jackfruits (*Artocarpus heterophyllus*) seeds. *Agriculture and Biology Journal of North America*. 1(5): 903-908.

Okilya, S., Mukisa, I. M. and Kaaya, A. N. (2010). Effect of solar drying on quality and acceptability of Jackfruit leather. *Electronic Journal of Environmental, Agricultural and Food Chemistry*. 9 (1): 101-111.

Pandian, T. and Sundarajan, K. (2014). Problems and prospects of Jackfruit cultivars- A study with reference to Tamil Nadu. *Asia Pacific Journal of Research*. 1(18): 63-71.

Patil, R, Joshi, G. D., Haldankar, P. M. and More, Mrinal (2011). Preparation of value added products from Jackfruit and organoleptic evaluation. *International Journal Processing and Postharvest Technology*. 2 (1): 44-51.

Punan, M. S., Rahman, A. S. B., Nor, L. M., Muda, P., Sapii, A. T., Yon, R. M. and Som, F. M. (2000). Establishment of a quality assurance system for minimally processed Jackfruit. Quality Assurance in Agriculture Produce. Edited by G. I. Johanson, Le Van To, Nguyen Duy Duc and M. C. Webb. ACIAR Proceeding 100. pp 115-122.

Rahim, M. A. and Quaddus, M. A. (2000). Characterization and grafting performance of different accessions of Jackfruit. M. Sc. Thesis, Bangladesh Agricultural University, Mymensingh, Bangladesh.

Rahman, M. A., Nahar, N., Mian, A. J. and Mosihuzzaman, M. (1999). Variation of carbohydrate composition of two forms of fruit from jack tree (*Artocarpus heterophyllus* L.) with maturity and climatic conditions. *Food Chemistry*. 52: 405-410.

Rahman, M., Miaruddin, M., Chowdhury, M. G. F., Khan, M. H. H. and Rahman, M. M. (2012). Preservation of Jackfruit (*Artocarpus heterophyllus*) by osmotic dehydration. *Bangladesh Journal of Agriculture Research*. 37 (1): 67-75.

Ranganna, S. (2007). Hand Book of Analysis and Quality Control for Fruits and Vegetables Products. Tata McGraw-Hill Publishing Company Limited: 1099.

Samaddar, H. N. (1985). Jackfruit. In: Bose, T.K. (Edited) Fruits of India: Tropical and Subtropical, Naya Prokash, Kolkata (W.B.) India. pp. 487-497.

Saravan Kumar, R. and Manimegalai, G. (2002). Storage stability of soya milk whey based Jackfruit juice blended R.T.S. beverages. *Processed Food Industry*. (Dec.): 42-47.

Seow, C. C. and Shanmugam, G. (1992). Storage stability of canned Jackfruit juice at tropical temperature. *Journal of Food Science and Technology*. 29(6): 371-74.

Sharma, J. P., Upadhyay, S., Chaturvedi, V. K. and Bharadwaj, T. (2014). Enhancing farm profitability through food processing and value addition. In: Shukla, J.P. Technologies for Sustainable Rural Development: Having Potential of Socioeconomic Upliftment (TSRD-2014). Published by allied Publisher Pvt. Ltd. India.p. 88.

Srivastava, A., Bisnoi, S. K. and Sarkar, P. K. (2017). Advance in value addition in Jackfruit (*Artocarpus heterophyllus* Lam.) for food and livelihood security of rural comunities of India. *The Asian Journal of Horticulture*. 12 (1): 160-164. DOI : 10.15740/HAS/TAJH/12.1/160- 164.

Swami, S. B., Thakor, N. J., Haldankar, P. M. and Kalse, S. B. (2012). Processing and value addition in Jackfruit. *International Journal of Processing and Postharvest Technology*. 3 (1): 142-146.

Tilak, B. (1995). Ghare karo shilpo garo. West Bengal State Book Board, Calcutta, India.

7

Jamun

Scientific Name: *Syzygium cumini*

Family: Myrtaceae

Edible portion: Mesocarp and Epicarp

When 70 per cent of Indian populations are engaged only in production activities of Agriculture, we need more and more entities and systems to add value. In doing so, we need to look at food wastages and their prevention, improvement in value addition of horticultural produces through adoptable processes, harnessing untapped food resources, utilizing by-products and assuring food quality and safety. All these have to be interlinked with extension of shelf life, which is also value addition. *Eugenia jambolana* is known as *Syzygium jambolanum* and *Syzygium cumini,* belongs to Myrtaceae family is an important indigenous minor fruit for commercial as well as medicinal value. Indian black berry, Black plum, Java plum, Jambul, Jamun, Jam, Kalajam, Phalani, Pharendra, are synonyms for Jamun (Singh, 1997; Morton and Miami, 1987; Zaman and Shariq, 1995; Swami *et al.,* 2012a).). This fruit is considered as underutilized fruit crop, which have high nutritional value and excellent processing qualities (Joshi 2001). Though there is a maximum availability of raw material or fruit harvested per year. It cannot be utilized or consumed and processed due to lack of processing techniques and technical knowhow. Jamun being a highly perishable and short shelf-life fruit it deteriorates at very faster rate if proper postharvest handling practices and processing techniques are not adopted (Deshmukh *et al.,* 2012). Considering the mass fruit production from the increasing plantation in coming days, proper postharvest handling practices for increasing its shelf life and processing techniques need to be explored. Its fruits are delicious and have great importance in folk medicine (Chopera, 1956). The Jamun is of wider interest for its medicinal applications than for its edible fruit. Different parts such as bark, fruit and seed possess medicinal and therapeutic values (Kirtikar *et al.,* 1990; Noomrio and Dahot, 1996). According to Hindu tradition Lord Rama subsisted in this fruit in the forest for 15 years during his exile from Ayodhya. Because of the above mention episode, many Hindu regard Jamun as **'fruit of the god'** especially in Gujarat. The tall evergreen trees are grown for the shade and windbreak on roadsides and avenues (Koley *et al.,* 2011).

People's growing interest in health is reflected in concern about cholesterol and low calorie foods, organic foods and antioxidant rich foods *etc.*; today's market is flooded with blended fruit based beverages. We have a habit of concentrating only on major fruit based products and ignoring the potential of under exploited fruit products in the context of value addition. Thus to give the variety to the taste of the consumer innovative items are always asked for. Under these circumstances, the processed product has been formulated to develop value added products, which are highly perishable and impossible to keep them for more than 24 hours under ambient condition *e.g.,* Jamun; some of them are not easy to eat out of hand and these seasonal in nature with short storage life even under low temperature condition, necessitate the processing of Jamun.

Importance and Use

The Jamun is of wider interest for its medicinal applications than for its edible fruit. The fruit is good source of iron, sugars, minerals, protein and carbohydrate. Fully ripened fruits are eaten as fresh fruit and can be processed into beverages like jelly, jam, squash, wine, vinegar and pickles (Ayar *et al.,* 2011). Fruits are used as an effective medicine against diabetes, heart and liver trouble (Singh, 2001). The leaves and the bark are used to control blood pressure and gingivitis (Joshi, 2001). Leaf extract of *Jamun* reduces the radiation induced DNA damage in the cultured human peripheral blood lymphocytes (Prince *et al.,* 2003). Therefore, the *Jamun* fruits are having high value in terms have therapeutic and nutrition. Jamun fruit is one of those, which contain a variety of important nutritional compositions. The

Composition of Jamun (*S. cumini*) Fruits

Constituents	*Content*	*Constituents*	*Content*
Moisture (per cent)	84.5-86.4	Pectin (per cent)	2.3-3.7
Protein (per cent)	0.53-0.64	Vitamin A (IU)	73-100
Fat (per cent)	0.10	Vitamin C (mg/100 g pulp)	30.3-40.7
Calorific value (per 100g)	83	Calcium (per cent)	0.02
TSS (^{o}Brix)	9.0-11.5	Phosphorus (per cent)	0.01
Total Sugar (per cent)	5.8-6.9	Iron (per cent)	0.1
Acidity (per cent)	2.1-2.5		

Source: Singh and Srivastava, 2000; Singh *et al.*, 1967.

Jamun fruit contains 83.7-85.8 g moisture, 0.7- 0.129 g protein, 0.15-0.3 g fat, 0.3-0.9 g crude fiber, 14 g carbohydrate, 0.32-0.4g ash, 8.3-15mg calcium, 35mg magnesium, 15-16.2mg phosphorus, 1.2-1.62 mg iron, 26.2 mg sodium, 55 mg potassium, 0.23 mg copper, 13 mg sulfur, 8 mg chlorine, 80 I.U. vitamin A, 0.008- 0.03 mg thiamine, 0.009- 0.01 mg riboflavin, 0.2-0.29 mg niacin, 5.7-18 mg ascorbic acid, 7 mg choline and 3 mg folic acid per 100 g of edible portion (Noomrio and Dahot, 1996; Swami *et al.,* 2012a). It has been well recognized that the consumption of large amounts of fresh fruits and vegetables can bring substantial health benefits.

Medicinal Value of Jamun

Jamun fruits are universally accepted to be very good for medicinal purposes especially for curing diabetes because of its effect on the pancreas (Joshi, 2001) and it quickly reduce sugar in urine (Zaman and Shariq, 1995). The fruit, its juice and the seed contain a biochemical called 'jamboline' which is believed to check the pathological conversion of starch into sugar in case of increased production of glucose. Beside, the Jamun fruit is an effective food remedy for bleeding piles and correcting liver disorders. Regular intake of fruits for 2-3 months could help with hemorrhoids. The fruit rind could be helpful in improving liver health. The real benefit of this is at the stage of IFG (Impaired fasting glucose). The cumulative and rational combination of certain herbs could offer blood sugar support. When Jambolan extract is used in conjunction with Gymnema, bitter melon, salacia, cinnamon and fenugreek the benefits are immense. Fruits reduce excessive salivation and when it's used as a supplement could help to enhance the insulin activity and sensitivity. Jamun fruit is used for the prevention of diarrhea, stomachache, astringent, diuresis and diabetes. The fresh Jamun juice is mixed with goat's milk and then given to children in diarrhea (Zaman and Shariq, 1995). It is also used for enlarged spleen, chronic diarrhea and urine retention. Water diluted juice is used as a gargle for sore throat and as a lotion for ringworm of the scalp. Its fruits are delicious and have great importance in folk medicine (Chopera, 1956). A decoction of bark is used in cases of asthma and bronchitis and are gargled or used as mouthwash for the astringent effect on mouth ulcerations, spongy gums, and stomatitis. Seeds contain an alkaloid jambosine and glycoside jambolin or antimellin, which halts the diastatic conversion of starch into sugar (Morton, 1987). Seed extract has been reported to lower blood pressure by 34.6 per cent due to the presence of ellagic acid (Morton, 1987). Seeds are also rich in flavonoids and are well know antioxidants (Ravi *et al.,* 2004). Seeds are fairly rich in protein, calcium and other minerals also (Ayyanar and Babu, 2012). Dried seeds powder three times a day at one gram each dose could help support blood sugar. Two grams of dried seeds powder could help with Polyuria of any cause. Both seeds and leaves improve uterine function. Seeds decoction with honey prevents thirst and fatigue due to physical strain. It is one of the best detoxifying agents when combined with certain herbs. Seeds decoction has antiseptic activity. The extract of Jamun seed lowers blood pressure more than 30 percent and this action is attributed to the ellagic acid content of the extract (Morton and Miami, 1987). If child is doing bet wetting then you should give 1-teaspoon ground seed with water to child. In case of dysentery (bloody) take 20 g powdered seeds and take with 1/2 cup of water twice daily. In case diarrhea take two soft leaves of Jamun and make paste of it. Now add some rock salt to it and make small pills and take twice daily with water. In case of conjunctivitis boil soft 20 leaves of Jamun and boil in 2-cup of water till it reduces to 100 ml. Cool down this and wash eyes with this water (decoction). In case of acidity drink 10 ml Jamun vinegar with some water it will give you relief (Bhowmik *et al.,* 2013). The tasty and pleasantly flavored Jamun fruit is mostly used for dessert purposes and it is very much liked by the people. The fruit is usually shaken with salt before eating. The Jamun fruit has sub acid spicy flavour. Apart from eating as fresh, it can also be used for making delicious beverages. Smaller fruits are used in beverage industry for being rich in acidity, tannins and anthocyanins. Its seed can be used as a concentrate for animals because it is rich in protein, carbohydrates, and calcium. The oil composition of Jamun has also been reported (Bose, 1985). According to Alam *et al.,* 2012 identify the putative anti-diabetic constituents from the *S. cumini* leaves and from the NMR data four different compounds, Lupeol, 12-oleanen-3-ol-3 β-acetate, Stigma-sterol, β-sitosterol were identified from n-hexane fraction of plant extract. These compounds have potential antidiabetic activities, which support the traditional use of the leaves as being remedy for treating diabetes.

Syzygium species have been reported to exhibit anti-diabetic (Nonaka *et al.,* 1992; Kumar *et al.,* 2008), antifungal (Park *et al.,* 2007; Ayoola *et al.,* 2008; Kiruthiga *et al.,* 2011), anti-inflammatory (Chaudhuri *et al.,* 1990), antibacterial (Shyamala and Vasantha, 2010), antioxidant (Nassar *et al.,* 2007), anti-hyperlipidemic (Modi *et al.,* 2010) and growth inhibitory effects against oral pathogens (Cai and Wu, 1996). *Syzygium* species are also found to posses free radicle scavenging activity (Rekha *et al.,* 2008; Zhang, 2009), anti-hyperglycemic activity (Rekha *et al.,* 2010), cytotoxic (Aisha *et al.,* 2011), anti-angiogenic (Aisha *et al.,* 2011), anti-nociceptive activity (Avila Pena, 2007).

The bark of the plant has various properties like astringent, refrigerant, carminative, diuretic, digestive, anti-helminthic, febrifuge, constipating, stomachic and antibacterial activity (Saravanan and Pari, 2008). The fruits and seeds are used to treat diabetes, pharyngitis, spleenopathy, urethrorrhea and ringworm infection (Saravanan and Pari, 2008). The leaves are antibacterial and used to strengthen the teeth and gums and extensively used to treat diabetes, constipation, leucorrhoea, stomachalgia, fever, gastropathy, strangury, dermopathy and to inhibit blood discharge in the feces (Ravi *et al.,* 2005; Sagrawat *et al.,* 2006; Gowri and Vasantha, 2010). The powders of stem bark mixed with curd are used for the treatment of blood dysentery (Pandey and Tripathi, 2011). The barks, leaves and seeds extracts of *S. Cumini* have also been reported to possess anti-inflammatory antibacterial and anti-diarrhea heal effects (Indira and Mohan, 1992; Chaudhuri *et al.,* 1990; Bhuiyan *et al.,* 1996). Powdered seeds are used as a remedy in diabetes and in menorrhagia (Sharma and Mehta, 1969). It has been also showed before that the leaf; bark, stem and pulp of *S. cumini* plants possess potent anti-diabetic activity (Chaudhary and Mukhopadhyay 2012; Kumar *et al.,* 2008; Leelavinothan and Saravanan, 2006; Farswana *et al.,* 2009; Bopp *et al.,* 2009).

The major phytoconstituents are reported to contain vitamin C, gallic acid, tannins, anthocyanins, includes cyanidin, petunidin, malvidinglucoside and other components (Martinez and Del V alle, 1981). Preliminary phytochemical analysis also showed the presence of phenols, terpenoids, tannins, saponins, phytosterols, carbohydrates, flavonoids, amino acids in stem bark of *S. cumini* (Kuncha *et al.,* 2012). Previous investigations also revealed the bark of *S. cumini* contains butulinic acid, β-sitosterol, friedelin, epi-friedelanol (Wealth of India, 1982). It also contains new esters of epi-friedelanol (eugenin), D-glucoside, kaempterol-3-O-glucoside, quercetin, myricetin, astragalin and gallic acid (Sen gupta and Das, 1965; Bhargava *et al.,* 1974; Williamsen *et al.,* 2002).

Phytochemicals Present in the Jamun Plant

Sl.No.	*Plant Part*	*Chemicals Present*
1	Seeds	Jambosine, gallic acid, ellagic acid, corilagin, 3,6-hexahydroxy diphenoylglucose, 1-galloylglucose, 3-galloylglucose, quercetin, β-sitoterol, 4,6 hexahydroxydiphenoylglucose, (Sagrawat *et al.,* 2006; Rastogi and Mehrotra, 1990).
2	Stem bark	Friedelin, friedelan-3-α-ol, betulinic acid, β-sitosterol, kaempferol, β-sitosterol-Dglucoside, gallic acid, ellagic acid, gallotannin and ellagitannin and myricetine (Sagrawat *et al.,* 2006; Rastogi and Mehrotra, 1990).
3	Flowers	Oleanolic acid, ellagic acids, isoquercetin, quercetin, kampferol and myricetin (Sagrawat *et al.,* 2006).
4	Fruit pulp	Anthocyanins, delphinidin, petunidin, malvidin-diglucosides (Sagrawat *et al.,* 2006; Li *et al.,* 2009; Veigas *et al.,* 2007).
5	Leaves	*dippa*-sitosterol, betulinic acid, mycaminose, crategolic (maslinic) acid, n-hepatcosane, n-nonacosane, n-hentriacontane, noctacosanol, n-triacontanol, n-dotricontanol, quercetin, myricetin, myricitrin and the flavonol glycosides myricetin 3-O-(4"-acetyl)-α-Lrhamnopyranosides (Sagrawat *et al.,* 2006 ; Ranjan *et al.,* 2011).
6	Essential oils	α-terpeneol, myrtenol, eucarvone, muurolol, α-myrtenal, 1, 8-cineole, geranyl acetone, α-cadinol and pinocarvone (Shafi *et al.,* 2002).

Source: Swami *et al.,* 2012.

Jamun fruit has been valued in Ayurveda and Unani system of medication for possessing variety of therapeutic properties (Chitnis *et al.,* 2012). It is rich in carbohydrates, minerals such as manganese, zinc, iron, calcium, sodium, potassium and vitamins (Ranjan *et al.,* 2011). The edible pulp contains various phytochemicals like Vitamin C, Vitamin A, riboflavin, choline, anthocyanin's and various other polyphenols (Choudhury and Mukhopadhyay, 2012). Jamun fruit has long been used for the treatment of diabetes prior to the discovery of insulin and possess various pharmacological properties such as antibacterial, antifungal, antiviral, anti-genotoxic, anti-inflammatory, anti-ulcer genic, cardio-protective, anti-allergic, anticancer, chemo-preventive, radio-protective, free radical scavenging, heat protective, anti-diarrheal and hypo-glycemic (Baliga *et al.,* 2011). Jamun fruits are processed to make jam, jellies, squash, vinegar and ice cream for its pleasing and attractive purple colour due to presence of anthocyanins. Thus, flavonoids present in the fruit play dual role as attractive natural colorants, which also add nutrition to the food and helps improve overall quality of food.

Therapeutic Value of Jamun Seed

1. Seeds contain an alkaloid Jambosine and glycoside jambolin or antimellin, which halts the diastatic conversion of starch into sugar (Morton, 1987).
2. Seed extract has been reported to lower blood pressure by 34.6 per cent due to the presence of ellagic acid (Morton, 1987).
3. Seeds are also rich in flavonoids and are well know antioxidants (Ravi *et al.,* 2005).
4. Seeds are fairly rich in protein, calcium and other minerals also (Ayyanar and Babu, 2012).

Uses of Jamun Seeds in Folk Medicine

Ethnic Group or Region	*Mode of Preparation, Administration and Ailments*
Lakher and Pawi in North east India	Juice of seed is applied externally on sores and ulcer. Powdered seeds mixed with sugar are given orally in the treatment of dysentery.
Malayalis in South India	Paste of seeds along with combination of leaves of bitter gourd and flowers of *Cassia auriculata* is used to treat diabetes
Traditional medical healers, Madagascar	Seeds are taken orally as an effective therapy for slow debilitating impacts of diabetes
Local population of A. P. (India)	Powder of shade dried seeds are taken orally thrice a day in treatment of diabetes

Source: Roy *et al.*, 2013.

Determination of Mean Values of Minerals in Various Products of Jamun Fruit (mg/kg) *cv.* V1

Sample	Cr	Cu	Mn	Zn	Fe	Mg	Ca	Na	K
Seed powder	5.2	22.7	13.2	6.9	9.4	98.5	214.9	2200	17051.01
Pulp powder	3.0	14.4	13.0	10.8	14.4	109.0	160.2	1667	13601.97
Jam	3.8	55.9	13.2	11.6	10.9	94.3	252.3	1886	21837.93
Squash	4.0	9.3	13.1	7.3	9.4	71.0	614.5	4673	9072.81
RTS	3.4	68.4	13.2	14.2	16.8	92.8	610.9	1949	8950.07
Fresh pulp	3.5	68.7	13.3	21.1	20.0	95.4	620.9	1417	17910.20

Source: Nawaz *et al.*, 2010.

The plant *Syzygium cumini* utilized in both antibacterial and antifungal activity of various extracts. Traditional herbal medicines must be placed in a position so that they support the benefits of modern science technology as global needs. The plant also contain phytomedicine, antioxidant and anti-depressive agents, therefore, *Syzygium cumini* drugs may have the possibility of use in modern medicine (Pareek *et al.,* 2015).

Harvesting and Yield

The seedling Jamun plants start bearing after 8 to 10 years of planting while grafted ones after four to five years. Jamun is a non-climacteric fruit. Hence quality of fruit can't be improved after harvesting. For this reason fruits are allowed to ripen on the tree. The fruits ripen in the month of May-July or with the onset of rain (Chowdhury and Ray, 2007; Koley *et al.,* 2011). The berry is oblong, ovoid and shinning crimson black (rich in anthocyanin pigment, an anti-oxidant) when fully ripe. Fruits of the grafted variety are large and deliciously sweet but slightly sour (Pathak and Pathak, 1993). Jamun is highly perishable and as per an estimate 20- 30 per cent of Jamun is lost during harvest and postharvest (Joshi, 2011). At ripening, colour of fruit changes from green to deep purple or black. Fruit should be harvested as soon as they ripe because they start dropping to the ground immediate after ripening. These are often picked from ground but the fruits are generally infected with soil borne microorganisms. Since most of the fruits are damaged due to impact injury once hit the ground, the common method of harvesting of Jamun is shake the tree and collect the fruits by holding as piece of cloth or canvas under the tree. Harvesting is also done by climbing on the tree and collecting fruits in a cloth bag on the shoulder of picker. The fruits cannot remain on the tree in the ripe stage, and start dropping to the ground immediately after ripening. As a result, a considerable amount of fruit is damaged and become unfit for fresh consumption. It is unfortunate that no proper technique for harvesting Jamun has yet been developed. Jamun fruits are generally harvested daily and send to market on the same day as they can't be kept for long time (Koley *et al.,* 2011). The average yield of fruit from a full-grown seedling Jamun tree is about 80 to 100 kg and from a grafted one 60 to 70 kg per year.

Storage

Jamun is highly perishable fruit. They cannot be stored for more than 2 days under ambient condition. Precooled fruit packed in perforated polyethylene bag can be stored up to 3 week at 8-10 ^{0}C with 85-90 per cent RH. Before packing, blemished fruit should be sorted out. Fruits are normally packed in bamboo basket and transported to local market. Very high quality fruits are packed in paper basket for long distance transportation. Fruits are pre-packed in leaf-cup cover with perforated polyethylene bag have little or no damage during handling and handling of fruit from market to home is also easier in this container (Koley *et al.,* 2011). According to Vandana *et al.* (2015) found that 1 per cent $CaCl_2$ treated Jamun fruits maintained low physiological loss (11.61 per cent), retained more firmness (135 g), low respiration rate (66.33 ml CO_2/Kg/hrs), soluble solids content (18.45 °Brix) and titratable acidity (0.82 per cent) and extended shelf life up to 14 days. The storage life of Jamun fruit is restricted to only 24 hrs at room temperature and 12 days at cool temp. *i.e.* 3 to 4 °C and 85 to 95 per cent RH (Ramanjan, 1985). Shukla (1979) observed that the storage life of Jamun fruit is 6 days at room temperature and 3 weeks at low temperature (9 ± 1 °C and RH 85 to 90 per cent) when pre-cooled fruits are kept in perforated polythene bags. Banik *et al.* (1986) reported that treatment of ripe Jamun fruits with 100 ppm citric acid or 100 ppm ethrel or keeping in polythene bags with 10 g permanganate-silica gel prolonged the shelf life in satisfactory condition up to 12 days. According to Ayar *et al.* (2011) studied the postharvest treatments on quality of Jamun fruits during storage and found that among all treatments the best result was in fruits treated with $CaCl_2$ 1.5 per cent with perforated polyethylene bag proved to be the best postharvest treatment than the rest of the treatments in respect of physiological loss in weight, spoilage loss, good balance between ascorbic acid and sugar content, TSS, pH and acidity.

Value Added Products from Jamun

Integrated Processing of Jamun

Mature ripe fruits of Jamun

↓

Juice extraction → Processed products (nectar, RTS, squash, syrup)

↓

Dried ← Pomace (seeds and peel) Residue (seeds and some parts of peel)

↓ ↓ ↓

Powder, tablet etc. Pomace extract Dried

↓ ↓

Beverage preparation (Nectar, RTS) Powder tablet

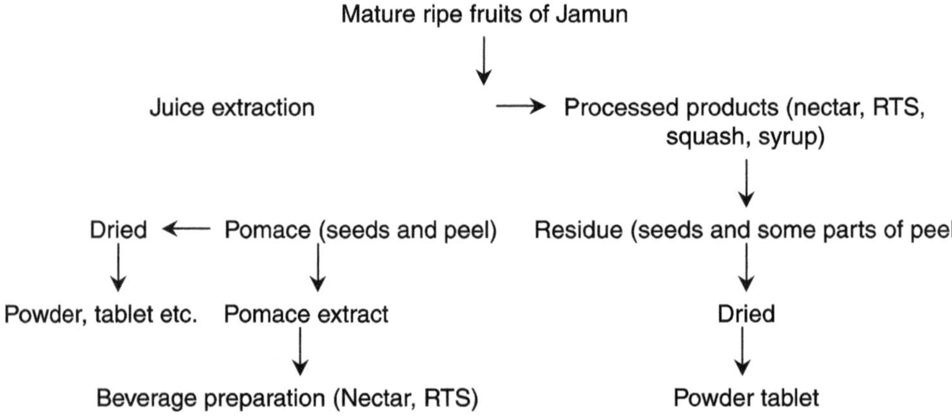

Fruit processing industry is not very viable in South Asian Countries due to non-utilization of waste. Integrated processing of Jamun will provide complete utilization of fruit so that nothing goes waste. Development of products from underutilized fruits like Jamun could be ideal for processed food market. The new nutritious natural and healthy processed foods are in great demand.

Jamun Juice

Sl.No.	Ingredients	Quantity	Sl.No.	Ingredients	**Quantity**
1	Jamun fruits	1.0 kg	3	Preservative (Sodium Benzoate)	150 ppm/ltr. of finished product
2	Water	½ ltr.			

Procedure

Good quality jambolan juice is excellent for sherbet (Benthall, 1946; Brown, 1920), syrup and "squash". Fresh ripe fruit wash properly and discord the spoiled fruits. Fruits add with water keep on heater for heating till just for soften (5 to 10 minutes at 140 ^{0}F). Pulping by pulper machine (pressing out the juice). Straining with the help of muslin clothes and collects the dark purple color juice. Filled in sterilized bottles with adding of sugar, water, citric acid and preservative and finally capping and sealing. Pasteurized at 80 ^{0}C for 20 minutes. Stored at refrigeration or room temperature (Lal *et al.,* 1960). Apart from pure Jamun juice, Jamun juice mixed with mango juice is very effective for quenching thirst for diabetic patients. Its juice being highly acidic in nature is not consumed as such but juice is stomachic, carminative and diuretic apart from having cooling and digestive properties. The attractive colour due to anthocyanin pigments is a major quality attributes in Jamun beverages. The anthocyanin pigments are destroyed at high storage temperature. Refrigerated storage has stabilizing effect on the anthocyanin of the fruit juices (Ponting *et al.,* 1960). A method of extraction of the Jamun juice has been standardized (Shukla *et al.,* 1991) for the preparation of ready-to-serve beverage (nectar). It has been found that the maximum yield of Jamun juice with a high level of the anthocyanin and other soluble constituents was obtained by grating the fruit, heating up to 70 ^{0}C and passing the heated mass through a basket press (Ramanjaneya, 1985; Reginold *et al.,* 2013). Roy *et al.* (1999) suggested that a maximum yield of Jamun juice with a high level of anthocyanin and other soluble constituents can be obtained by grating fruits, heating it to 60 ^{0}C and passing the grated mass through a basket press. Khurdiya and Roy (1985) have been studied the quality of Jamun juice and nectar during storage at different temperature including the cool chamber developed by Roy and Khurdiya (1982) with special reference to stability of colour. They observed that juice was acceptable up to 8 months storage at room temperature and 12 months at cool temperature, whereas the nectar can be stored for 10 months and 12 months at room and at cool temperature respectively. Jamun juice is used to cure the diabetes disease effectively. It is also useful against bleeding piles, correcting liver disorders, jaundice, kidney stone, asthma, blood pressure, (Wealth of India 1954; Joshi 2001; Deshmukh *et al.,* 2012). Deshmukh *et al.* (2012) stored Jamun juice with different combination of sodium benzoate at room and refrigerated condition and it was found that they were safer as compare to pure Jamun juice at room temperature. Juice of Jamun and mango if mixed in equal quantity is good in quenching thirst of diabetic patients. According to Kapoor and Ranote (2015) studied the antioxidant potential and quality of blended pear Jamun juice in which Jamun pulp were used in the ration of 5, 10, 15, 20 and 25 per cent and observed that the 20:80 (Jamun pulp: pear pulp) was best among the all treatments in respect of ascorbic acid, anthocyanin, total phenol and antioxidants.

Extraction of Juice

Commonly juice is extracted from fresh healthy fruits by crushing and pressing technology. But, there are different types of machine has been developed such as Screw type juice extractor and fruit pulper. These are mostly used for juice extraction but extraction of juice varies from type of fruit to fruit, because of differences in their structure and composition.

Clarification of Juice

Prasad and Mali (2000) reported that after the extraction, juice kept for 3-4 hours to down the coarse tissue particles. The supernatant solution was siphoned of leaving the coarse particles.

Bottling and Storage

The Jamun products (RTS, squash and syrup) stored in colorless glass bottles were acceptable even after six months at room temperature (Kannan and Thirumaran, 2001; Gavande *et al.,* 1995). Conventional thermal processing of fruit juices remains the most widely adopted technology for shelf life extension and preservation of fruit juice. However, consumer demand for nutritious foods, which are minimally and naturally processed, has led to interest in non-thermal technologies (Shaheer *et al.,* 2014). Jebitta and Alwin (2016) study the different types of drier (Sun drying, Cross flow drying, Vacuum shelf drying and Freeze drying) for Jamun pulp, and found that the Freeze dried powder of Jamun pulp powder has a good quality, flow ability, colour and solubility on comparing the other type of driers.

Flow-Sheet for Extraction of Juice

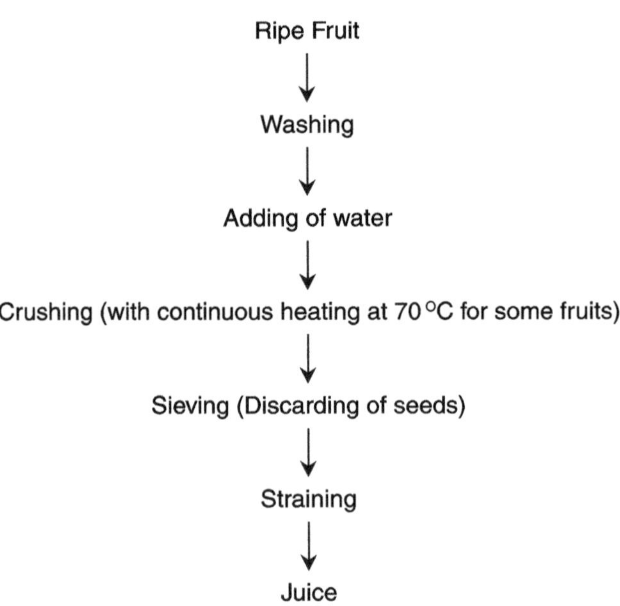

Utilization of Jamun Juice

Jamun juice can be processed into different types of beverages like RTS, nectar, syrup *etc.* A method of concentration of Jamun juice on lab scale has been standardized by Ramanjan (1985), which can be used by the beverage industries.

Utilization of Pomace

During extraction of Jamun juice a large amount of pomace remains as waste. It contains a considerable amount of anthocyanin's, tannins and sugars. A pomace extract can be obtained by mixing water with pomace in standardized ratio and this can be used for making fruit beverages.

Jamun RTS (Ready to Serve)

Sl.No.	*Ingredients*	*Quantity*	*Sl.No.*	*Ingredients*	*Quantity*
1	Juice	120 ml	4	Water	850 ml
2	Sugar	120 g	5	Preservative (SB)	150 ppm/ltr. of finished product
3	Citric acid	2.5 g			

Preparation of Syrup

Boil the sugar with water in stainless steel utensil, and add the citric acid to the boiling syrup for clarification. Strain the syrup and cool it.

Add the syrup little by little to the mixed with pulp and stirring, and get final RTS. Filled in sterilized bottles with adding of preservative, capping and sealing. Pasteurized at 80^{0}C for 20 minutes. Stored at refrigeration or room temperature. A ready to serve beverages (nectar) of Jamun was prepared with 25 per cent juice, 18 ^{0}Brix and 0.6 per cent acidity, it was found to be highly acceptable by Khurdiya and Roy (1985).

Technological Flow-Chart of Preparation of RTS

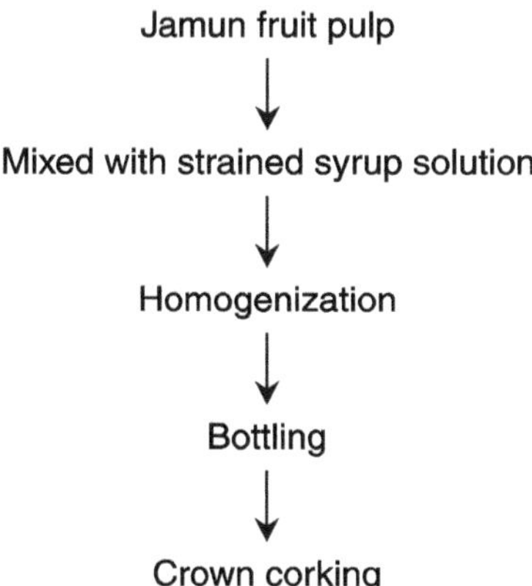

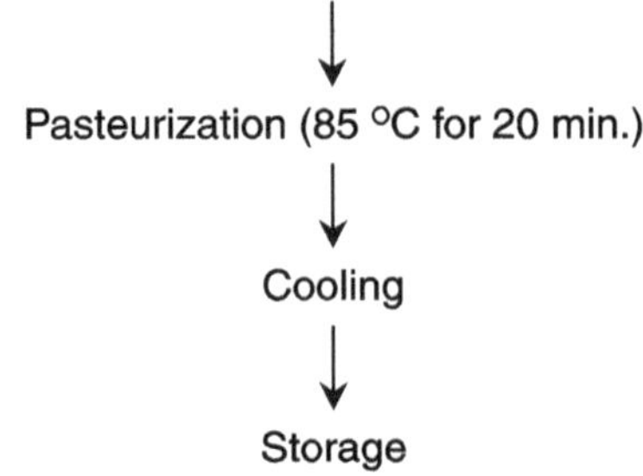

Source: Koley *et al.*, 2011; Srivastava and Kumar 2003.

Name of Products	*FPO Specification, 1955 (Minimum ingredients)*		*Other Ingredients*	
	Juice (per cent)	*TSS (°Brix)*	*Juice (per cent)*	*TSS (°Brix)*
Ready to serve	10	10	10-12	10-14
Squash	25	40	25-30	40-50
Nectar	20	15	20-25	15-20

Jamun Squash

Jamun squash is a very refreshing drink in the summer season. A little quantity of fruit syrup is much useful for curing diarrhea. Jamun syrup is beneficial in diarrhoea.

Sl.No.	*Ingredients*	*Quantity*	*Sl.No.*	*Ingredients*	*Quantity*
1	Juice	1000 ml	4	Water	2000 ml
2	Sugar	1860 g	5	Preservative (SB)	1 g per ltr. of finished product
3	Citric acid	42.0 g			

Technological Flow-Chart of Preparation of Squash

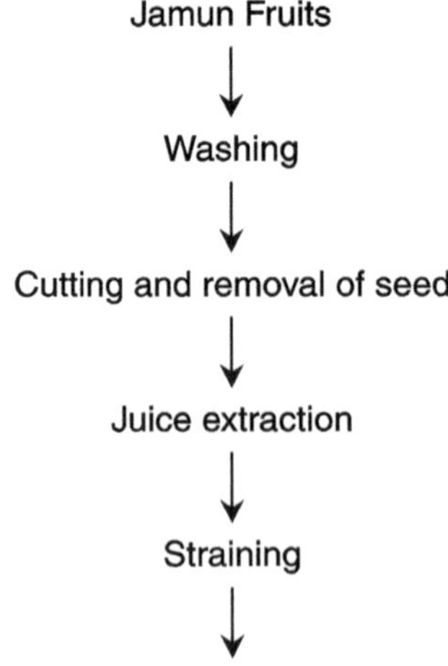

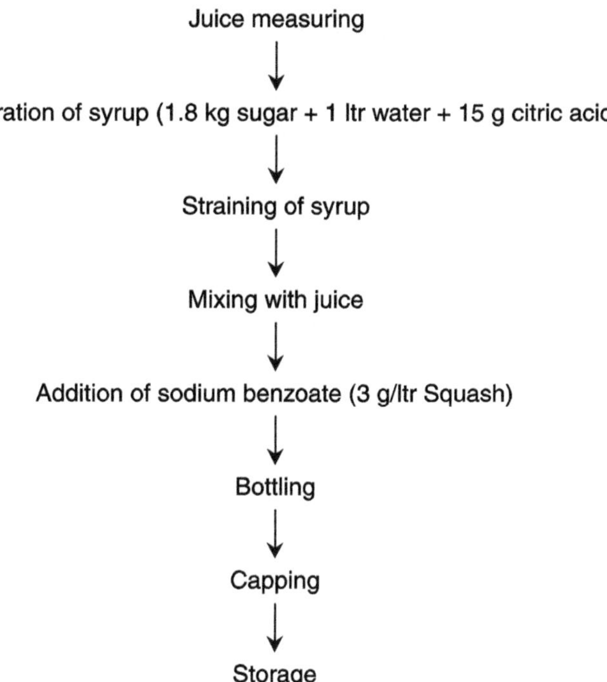

Source: Koley *et al.*, 2011.

Litchi and Jamun Blended Squash

Firstly collect the juice from Litchi and Jamun, and then Juice from both the fruits is blended together in a requisite proportion. Rest of the method is same as Jamun squash.

Sl.No.	*Ingredients*	*Quantity*	*Sl.No.*	*Ingredients*	*Quantity*
1	Jamun juice	125 ml	4	Citric acid	10.5 g
2	Litchi juice	125 ml	5	Water	450 ml
3	Sugar	450 g	6	Preservative (SB)*	1.0 g/litr

SB: Sodium Benzoate.

Jamun Wine

Ripe Jamun contain approximately 83 per cent water with almost solids 14 per cent containing a mixture of fermentable sugar. The pulp of Jamun contains appreciable amounts of fermentable sugar, which can be used for alcoholic fermentation (Kraus, 2003). According to Chowdhry and Ray, 2007 reported that a red wine from anthocyanin-rich tropical Jamun fruit having medicinal (anti-diabetic and curing bleeding piles) properties was prepared by fermentation using wine yeast (*Saccharomyees cerevisiae*) and the quality attributes compared with commercial grape red wines. The wine was sparkling red in colour, acidic in taste [titratable acidity (1.11 ± 0.07 g tartaric acid. 100 ml^{-1}), high tannin (1.7 ± 0.15 mg. 100 ml^{-1}) and low alcohol (6 per cent) concentration. Though sensory evaluation rated the Jamun wine quite acceptable as an alcoholic beverage, significant differences ($P < 0.05$) exist between the Jamun wine and the commercial grape wine particularly in taste, flavour and after taste probably due to the high tannin content in the Jamun wine. According to Patil *et al.* (2012) found that the wine which had 8.22 per cent alcohol obtained by fermentation with 10 per cent yeast level, 8.18 per cent TSS was the best acceptable when compared with others treatment like 6.62 per cent alcohol obtained by fermentation with 5 per cent yeast level, 7.88 per cent TSS. This wine was found to be the best by the panel members with respect to colour, taste, astringency, and flavor also. Brandy and distilled liquor called "Jambava" have also been made from the fermented fruit (USDA, 1912). Jambolan vinegar, extensively made through out India (USDA, 1912), is an attractive, clear purple, with a pleasant aroma and mild flavor. The vinegar prepared from the slightly unripe fruit is stomachic, carminative and diuretic, apart from having cooling and digestive properties.

Technological Flow-Chart for making Jamun Wine

Jamun Fruits

↓

Washing and cleaning by tap water and dip in 5 per cent NaCl salt solution for 72 hrs.

↓

Separate the seed from the fruit

↓

Crushed the fruit pulp with water (1:1 ratio) in a mixture-cum-grinder

↓

Pressing and extraction of juice; add SMS (100 μg/ml)

↓

Adjust TSS to 17 °Brix with cane sugar. Acidifying the must to pH 4.5 using 1N acetic acid

↓

Inoculation with wine yeast starter culture (use 24-48 hrs.
old starter culture at 2 per cent v/v)

↓

Fermentation (at 32± 2 °C for 6 days)

↓

Racking and decantationCarrying out first racking when the Brix reaches 2-3 °. Two to three racking at 15 days intervals if sedimentation persists

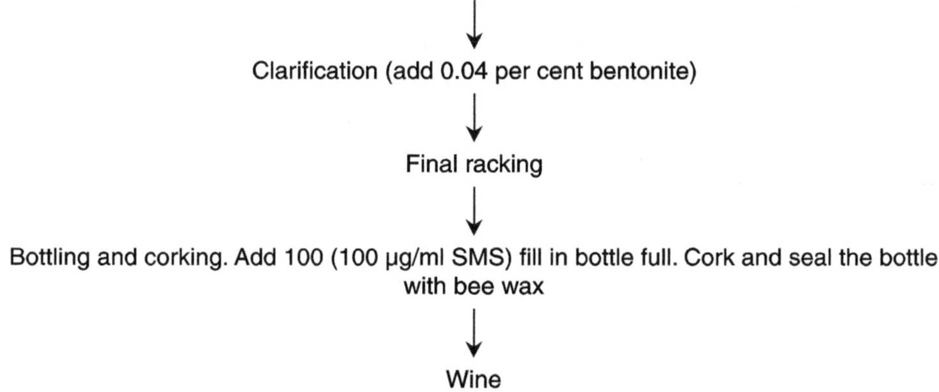

Source: Chowdhury and Ray, 2007.

Seeds

The seed is used as an alternative natural healing system in the Ayurvedic, Unani and Chinese medicines (Chaudhary and Mukhopadhyay, 2012). Older reports from Indian medical journals suggest jambul seed and bark can be beneficial in humans with diabetes (Sepaha and Bose, 1956). According to Thorat and Khemnar (2014) studied on Jamun seed fortified cookies by fortifying Jamun seed powder at different level (20, 30 and 40 per cent). Among the treatments 30 percent Jamun seed powder was recorded the highest scores for organoleptic parameters such as, colour, appearance, flavour, crispiness, taste and overall acceptability even up to 30 days of storage. Jamun Seed powder can be recommended to the patients those are suffering from diabetics because it contents more amount of jamboline, which lower sugar level in the blood. Jamun seed powder used as remedy for madhumeha, udakameha, atisara, raktpitta. Jamun fruit seeds and pulp has been reported to serve various purposes in diabetic patients, such as lowering blood glucose levels and delaying diabetic complications including neuropathy and cataracts (Sagrawat *et al.,* 2006; Helmstadter, 2008). Ferulic acid is a hydroxycinnamic acid, reported to possess an anticancer, antioxidant, and anti-aging potentials. Studies have shown that ferulic acid can decrease blood glucose levels and levels of cholesterol and triglycerides. Estimation of ferulic acid from ethyl acetate herbal extract of *Syzygium cumini* seed powder have anti-diabetic, anti-inflammatory, hepato-protective, anti-hyperlipidemic, diuretic and antibacterial activities (Shah *et al.,* 2012). Jebitta and Allwin (2016) study on different drying methods for Jamun pulp like using Sun Drying, Cross flow drying, Vacuum shelf Drying and Freeze drying and found that the Freeze Dried powder of Jamun Pulp powder has a good quality, flow ability, colour and solubility on comparing the other type of driers.

Jamun Mango Blended Jam

Jamun fruit, which is enriched in nutritive as well as medicinal value, was processed into jam and ascertain their storage stability; the jam was packed in glass, plastic and poly packs and kept at room temperature and it was safe nutritionally, microbiologically and organoleptically for a period of six months (Sood, 2015). The Jamun fruits were washed crushed for the extraction of pulp and on the other hand mango fruit was washed, peeled and passed through the pulper for obtaining the pulp. The pulp so obtained was passed through stainless steel strainer, homogenized followed by heating at 85 ^{0}C for 30 second.

Preparation of Jamun-Mango Blended Jam

Jam was prepared using 55 percent of blended Jamun: mango pulps as per the procedure of Prasad and Mali (2006). Jamun and mango fruits pulp were blended in the ratios of 55: 00: Jamun: mango pulp as control, 50: 05: Jamun: mango, 45: 10: Jamun: mango, 40: 15: Jamun: mango and 35: 20: Jamun: mango. Blended pulp was cooked along with the desired quantities of sugar and citric acid with constant stirring on a uniform flame, till the total soluble solids and acidity reached to 68 ^{0}Brix and 0.4 percent, respectively. The jam was filled hot in pre-sterilized glass bottles, cooled, sealed and kept upside down for some time then stored at room temperature after proper labeling. According to Sharma (2014) found that combination of Jamun mango on quality and storability of jam shows that 45:10, Jamun: mango combination was the best and maximum anthocyanin content retain as well as it scored maximum points for aroma, body, colour and overall acceptability after 6 month of storage.

According to Kapoor *et al.* (2015a) study the supplementation of wheat flour with *Jamun* pulp to improve nutritional and antioxidant status of *chapatti* in different ratio and 10 per cent pulp content of Jamun for *chapatti* making was best in anthocyanin, total phenol, antioxidants and organoleptically. By Kapoor *et al.* (2015) addition of 'Jamun' fruit powder in whole-wheat flour, help improved antioxidant activity, total phenolic content and anthocyanin content in 'chapattis'. Among the drying techniques, freeze-drying was found to better retain bioactive components than hot air drying.

Technological Flow-Chart of Preparation of Jam

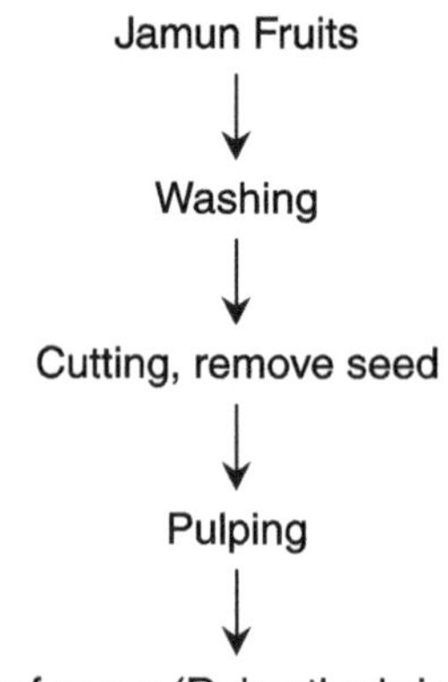

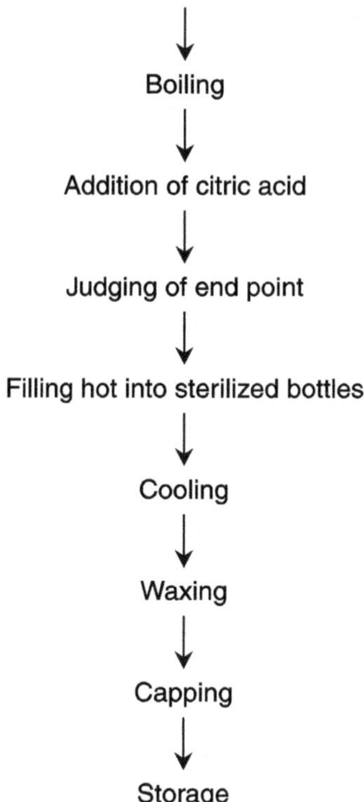

Source: Koley *et al.*, 2011.

Jamun Enriched Shrikhand (JES)

Keeping in view the nutritional, medicinal and delicious significances of Jamun fruit, develop value added JES. This precious fruit should be explored the nutritional and functional properties of Shrikhand which provide promising health benefits. For preparation of JES required materials are concentration of Jamun pulp powder and sugar in the range of 7- 15 per cent and 25- 35 per cent, respectively. According to the results of the desirability (0.83) function the combination of a 12.78 per cent Jamun pulp powder and 30.66 per cent sugar concentration provided the best results in relation to the average score of colour, flavour, sweetness, body and texture and overall acceptability of the JES were 8.25, 7.98, 8.41, 8.69 and 8.3 respectively (Singh and Paswan, 2015).

Flow-Chart of Preparation of Jamun Enriched Shrikhand (JES)

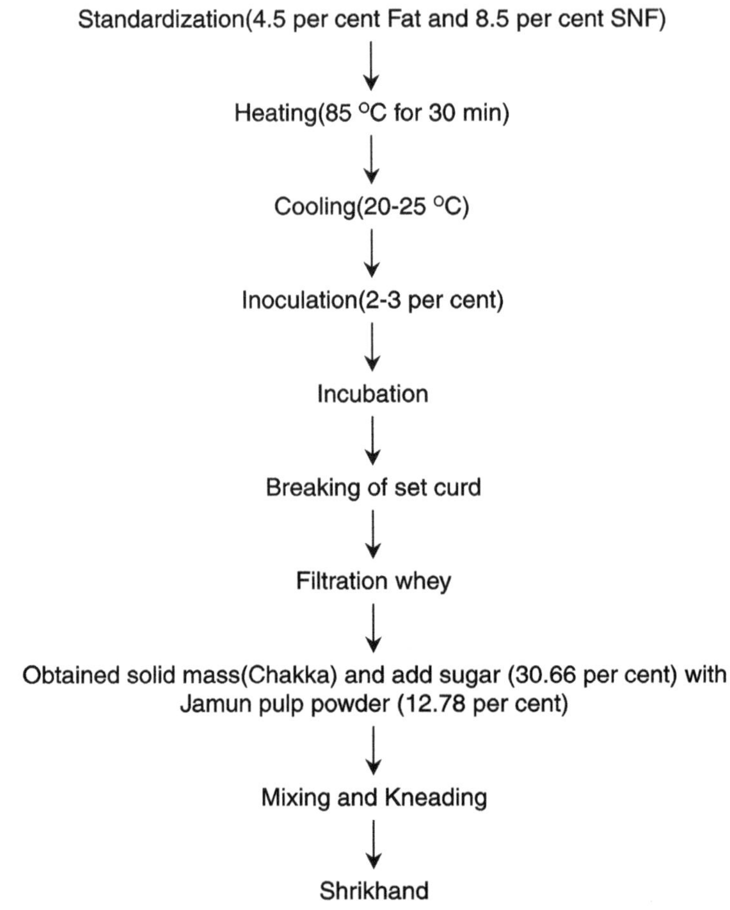

Source: Singh and Paswan, 2015.

REFERENCES

Aisha, A. F. A., Nassar, Z. D., Siddiqui, M. J., Abu Salah, K. M., Alrokayan, S. A., Ismail, Z. and Abdul Majid, A. M. S. (2011). Evaluation of antiangiogenic, cytotoxic and antioxidant effects of *Syzygium aromaticum* L. extracts. *Asian Journal of Biological Sciences*. 4 (3): 282-290.

Alam, M. R., Rahman, A. B., Moniruzzaman, M., Kadir, M. F., Hque, M. A., Alvi, M. R. U. H. and Ratan, M. (2012). Evaluation of antidiabetic phytochemicals in *Syzygium cumini* (L.) Skeels (Family: Myrtaceae). *Journal of Applied Pharmaceutical Science*. 2 (10): 94-98.

Avila Pena, D., Pena, N., Quintero, L. and Suarez Roca, H. (2007). Antinociceptive activity of *Syzygium jambos* leaves extract on rats. *Journal of Ethnopharmacology*. 112 (2): 380-385.

Ayar, Kiran, Singh, Sanjay and Chavda, J. C. (2011). Effect of postharvest treatments on quality of *Jamun* (*Syzygium cuminii* Skeels) fruits during storage. *Asian Journal of Horticulture*. 6 (3) : 297- 299.

Ayoola, G. A., Lawore, F. M., Adelowotan, T., Aibinu, I. E., Adenipekun, E., Coker H. A. B. and Odugbemi, T. O. (2008). Chemical analysis and antimicrobial activity of the essential oil of (*Syzygium aromaticum* (Clove). *African Journal of Microbiology Research*. 2: 162-166.

Ayyanar, M. and Babu, P. S. (2012). *Syzygium cumini* (L.) Skeels: A review of its phytochemical constituents and traditional uses. *Asian Pacific Journal of Biomedicine*. 2(3): 240–246.

Baliga M. S., Bhat H. P., Baliga B. R. V., Wilson R. and Palatty P. L. (2011). Phytochemistry, Traditional uses and pharmacology of *Eugenia jambolana* Lam. (black plum): A Review. *Food Research International*. 44: 1776-1789.

Banik, D., Dhua, R. S., Ghosh S. K. and Sen, S. K. (1986). Extension of storage life of Jamun (*Eugenia jambolana* L.). *South Indian Horticulture*. 34: 293-299.

Benthall, A. (1946). "Trees of Calcutta and Its Neigh Borhood. Thacker," Spink and Co. Ltd., Calcutta.

Bhargava K. K., Dayar R. and Seshadri T. R. (1974) Chemical components of Eugenia jambolona stem bark. *Current Science*. 43(20): 645-646.

Bhowmik, D., Gopinath, S., Kumar, B. P., Duraivel, S., Arvind, G. and Kumar K. P. S. (2013). Traditional and medicinal uses of Indian Black Berry. *Journal of Pharmacognosy and Phytochemistry*. 3 (5): 36-41.

Bhuyan, M. A., Mia, M. Y. and Rashid, M. A. (1996). Antibacterial principles of the seed of *Eugenia jambolana*. *Bangladesh Journal of Botany*. 25 : 239–241.

Bopp, A., Bona, K. S., Belle, L. P., Moresco, R. N. and Moretto, M. B. (2009). *Syzygium cumini* inhibits adenosine deaminase activity and reduces glucose levels in hyperglycemic patients. *Fundamental and Clinical Pharmacology*. 23: 501–507.

Bose, T. K. (1985). Fruits of India, Tropical and Subtropical. Naya Prokash, Calcutta, India. pp. 581-583.

Brown, W. H. (1920). "Wild Food Plants of the Philippines," *Bulletin* 21, Department of Agriculture and Natural Resources, Bureau of Printing, Manila.

Cai, L. and Wu, C. D. (1996). Compounds from *Syzygium aromaticum* possessing growth inhibitory activity against oral pathogens. *Journal of Natural Products*. 59 : 987-990.

Chaoudhury, P. and Ray, R. C. (2007). Fermentation of Jamun (*Syzgium cumini* L.) Fruits to Form Red Wine. *Asian Food Journal*. 14 (1): 15-23.

Chaudhary, B. and Mukhopadhyay, K. (2012). *Syzygium cumini* (L.) skeels: a potential source of nutra-ceuticals. *International Journal of Pharmacy and Biological Sciences*. 2 (1): 46-53

Chaudhuri, A. K. N., Pal, S., Gomes, A. and Bhattacharya, S. (1990). Anti-inflammatory and related actions of *Syzygium cuminii* seed extract. *Phytotherapy Research.* 4 : 5-10.

Chitnis K. S., Palekar S. B., Koppar D. R. and Mestry D. Y. (2012). Evaluation of syzygium cumini linn, Seed formulations available in the market using spectrophotometric and chromatographic techniques. *International Journal of Pharmaceutical Sciences and Research.* 3(2): 556-560.

Chopera, I. C. (1956). Glossary of Indian medicinal plants. ICSIR, Publication, New Dehli, India, pp: 44.

Deshmukh, L. S., Raut, V. U., Bhusari, R. B. and Khapare, L. S. (2012). Studies on sensory qualities during processing and storability of Jamun juice. *International Journal of Processing and Postharvest Technology.* 3 (20: 200-202.

Farswana, M., Mazumder, P. and Parcha, V. (2009). Modulatory effect of an isolated compound from *Syzygium cumini* seeds on biochemical parameters of diabetes in rats. *International Journal of Green Pharmacy.* 3 : 128-133.

Gavande, U. K., Joshi G. D. and Wasker, D. P. (1995). Storage of Jamun (*Syzygium cuminii*) fruit product. *ASEAN Food Journal.* 10 (2): 54-56.

Gowri, S. S. and Vasantha, K. (2010). Phytochemical Screening and Antibacterial Activity of *Syzygium cumini* (L.) (Myrtaceae) Leaves Extracts. *International Journal of Pharm Tech Research.* 2 :1569- 1573.

Helmstadter. (2008). "Syzygium *cumini* (L.) Skeels (Myrtaceae) Against Diabetes: 125 Years of Research". *Pharmazie.* 63 (2): 91-101.

Indira, G. and Mohan, R. M. (1992). National Institute of Nutrition, Indian Council of Medical Research, Hyderabad, India, pp. 34-37.

Jebitta, S. R. and Alwin, S. I. J. (2016). Effect of drying methods on the physical properties of Jamun pulp powder. *Indian Journal of Science.* 23 (77): 50-59.

Joshi, P. V. (2011). Postharvest handling and marketing of Jamun (*Syzygium cumini*) in Sindhudurg District of Maharashtra state. *International Journal of Communication and Business Management.* 4 (2): 249-253.

Joshi, S. G. (2001). *Medicinal plants,* New Delhi, Oxford and IBH Publishing Co.

Kannan, S. and Thirumaran, A. S. (2001). Storage life of Jamun products. *Processed Food Industry.* 5 (1): 18-19.

Kapoor, S. and Ranote, P. S. (2015). Antioxidant potential and quality of blended pear-Jamun (*Syzigium cumini* L.) juice. *International Research Journal of Biological Sciences.* 4 (4): 30-37.

Kapoor, S., Ranote, P. S. and Sharma, S. (2015). Bioactive Components and Quality Assessment of Jamun (*Syzygium cumini* L.) Powder Supplemented Chapatti. *Indian Journal of Science and Technology.* 8 (3): 287–295.

Kapoor, S., Ranote, P. S. and Sharma, S. (2015a). Antioxidant potentials and quality aspects of Jamun (*Syzygium cumini* L.) supplemented unleavened flat bread (*Indian chapatti*). *Journal of Applied and Natural Science.* 7 (1): 309-315.

Khurdiya, D.S. and Roy S. K. (1985). Processing of Jamun fruit into ready-to-serve beverages. *Journal of Food Science and Technology*. 22: 27.

Kirtikar, K. K., Basu, B. D. and I. C. S. (1990). Indian medicinal plants. Eds by O Blatter, J.F. Calus and K.S, Mhasker, 2, Ed. vol.2, 1052 pp.

Kiruthiga, K., Saranya, J., Eganathan, P., Sujanapal, P. and Parida, A. (2011). Chemical composition, antimicrobial, antioxidant and anticancer activity of leaves of *Syzygium benthamianum* (Wight ex Duthie) Gamble. *Journal of Biologically Active Products from Nature*; 1 :2738.

Koley, T. K., Barman, K. and Asrey, R. (2011). Nutraceutical Properties of Jamun (*Syzygium cumini* L.) and its Processed Products. *Indian Food Industry*. 30 (3): 43-46.

Kraus, E. C. (2003). Home wine and beer making supplies: http//www.eckraus.com.

Kumar, A., Ilavarasan, R., Jayachandran, T., Deecaraman, M., Aravindhan, P., Padmanabhan, N. and Krishan, M. R. V. (2008). Anti-diabetic activity of *Syzygium cumini* and its isolated compound against streptozotocin-induced diabetic rats. *Journal of Medicinal Plants Research*. 2: 246-249.

Lal, G., Siddappa, G. S. and Tandon, G. L. (1960). Preservation of Fruits and Vegetables. Indian Council of Agricultural Research, New Delhi.

Leelavinothan, P. and Saravanan, G. (2006). Effects of *Syzygium Cumini* bark on blood glucose, plasma insulin and C-peptide in streptozotocin- induced diabetic rats. *International Journal of Endocrinology and Metabolism*. 4 : 96-105.

Li, L., Adams, L. and Chen, S. (2009). "*Eugenia jambolana* Lam. Berry Extract Inhibits Growth and Induces Apoptosis of Human Breast Cancer but Not Non-Tumorigenic Breast Cells". *Journal of Agriculture and Food Chemistry*. 57 (3): 826-831.

Modi Dikshit, C., Rachh, P. R., Nayak, B. S., Shah, B. N., Modi, K. P., Patel, N. M. and Patel, J. K. ((2010). Antihyperlipidemic acitivity of *Syzygium cumini* Linn. Seed extract on high cholesterol (fed diet rats). *International Journal of Pharmaceutical Sciences*. 1: 330-332.

Morton, J. F. and Miami, FL. (1987). Jambolan. In: Fruits of warm climates. Pp. 375-378.

Nassar, M. I., Gaara, A. H., Ghorab, A. H. E., Farrag, A. R. H., Shen, H., Huq, E. and Mabry, T. J. ((2007). Chemical constituents of clove (*Syzygium aromaticum,* Fam. Myrtaceae) and their (antioxidant activity, Revista Latinoamericana de Quimica, 35 : 47-57.

Nawaz, M. S., Sheikh, S. A., Nizamani, S. M., Bhanger, M. I. and Afridi, I. (2010). Determination of mineral elements in Jamun fruits (*Eugenia jambolana*) product. *Pakistan Journal of Food Science.* 20 (1-4):1-7.

Nonaka, G., Aiko, Y., Aritake, K., Nishioka, I. (1992). Tannins and related compounds. CXIX: (Samarangenins A and B, novel proanthocyanidins with doubly bonded structures, from *Syzygium* (*samarangens* and *Syzygium aqueum*). *Chemical and Pharmaceutical Bulletin.* 40: 2671-2673.

Noomrio MH, Umar Dahot M (1996). Nutritive value of *Eugenia jambosa* fruit. *Journal of Islamic Academic of Sciences.* 9 (1): 9-12.

Pandey, A. K. and Tripathi, N. N. (2011). Aromatic plants of Gorakhpur division: their antimycotic properties and medicinal value. *International Journal of Pharmaceutical Sciences Review and Research.* 7 (2): 142-147.

Pareek, A., Meena, R.K. and Yadav, B. (2015). Antimicrobial activity of Syzygium cumini. Indian Journal of Applied Research. 5 (9): 64-66.

Park, M. J., Gwak, K. S., Yang, I., Choi, W. S., Jo, H. J. and Chang, J. W. (2007). Antifungal activities of the essential oils of *Syzygium aromaticum* (L.) Mer. Et Perry and *Leptospermum petersonii* Bailey and their constituents against various dermatophytes. *Journal of Microbiology.* (45: 460-465.

Pathak, R. Y. and Pathak, R. A. (1993). Improvement of minor fruits. *In:* Chadha, K. L. and Pareek, O. P. (Eds). Advances in Horticulture, Vol. 1, pp. 407- 422. New Delhi: Malhotra Publishing House.

Patil. S. S., Thorat, R. M. and Rajasekaran, P. (2012). Utilization of Jamun fruit (*Syzygium cumini*) for production of red wine. *Journal of Advanced Laboratory Research in Biology.* 3 (3): 200-203.

Ponting, J. D., Samshuck, D. W. and Brekke, J. E. (1960). Colour and deterioration in grape and berry juices and concentrates. *Food Research.* 25 : 471.

Prasad, R. N. and Mali, P. C. (2000). Changes in physico-chemical characteristics of pomegranate squash during storage. *Indian Journal of Horticulture.* 5 (1): 18-21.

Prasad, R. N. and Mali, P. C. (2006). Changes in physico-chemical characteristics of ber jam during storage. *Indian Journal of Horticulture.* 63 (1): 86-87.

Prince, P. S. M., Kamalakkanan, N. and Menon, V. P. (2003). Syzygium Cuminii Skeels seed extracts reduce tissue damage in diabetic rat brain. *Journal of Anthropology.* 84 : 205-209.

Ramanjan, Y. K. H. (1985). Studies on some aspects of Jamun fruit and its processing. Ph.D. Thesis, Indian Agricultural Research Institute, New Delhi (India).

Ranjan, A., Jaiswal, A. and Raja, B. (2011). "Enhancement of *Syzygium cumini* (Indian Jamun) Active Constituents by Ultra-Violent (UV) Irradiation Method." *Scientific Research and Essays,* 6 (12): 2457-2464.

Rastogi, R. and Mehrotra, B. (1990). "Compendium of Indian Medicinal Plants," Central Drug Research Institute, Lucknow, 1: 388-389.

Ravi, K., Rajasekaran, S. and Subramanian, S. (2005). Antihyperlipidemic effect of *Eugenia jambolana* seed kernel on streptozotocin-induced diabetes in rats. *Food and Chemical Toxicology*. 43 : 1433-1439.

Reginold, J. S., Ramanathan, M. and Praveen, S. (2013). Phytochemical Investigation of Cross Flow Dried Jamun (*Syzygium Cumini*) Pulp Powder. *International Journal of Engineering Research and Technology*. 2 (12): 992-994.

Rekha, N., Balaji, R. and Deecaraman, M. (2008). Effect of aqueous extract of Syzygium cumini pulp on antioxidant defense system in Streptozotocin induced diabetic rat. *Iranian Journal of Pharmacology and Therapeutics*. 7 (2): 137-145.

Rekha, N., Balaji, R. and Deecaraman, M. (2010). Antihyperglycemic and antihyperlipidemic effects of extracts of the pulp of *Syzygium cumini* and bark of Cinnamon zeylanicum in streptozotocin- induced diabetic rats. *Journal of Applied Bioscience*. 28 : 1718-1730.

Roy, S. K., Saran, S. and Mishra, V. (2013). Integrated Processing of *Jamun* (*Syzygium cumini* Skeels) Fruit For Value Addition and Assesment of its Impact on Health and Nutrition. Workshop on "Valorisation of traditional processing of indigenous underutilized fruits". Organized under the project "International network on preserving safety and nutrition of indigenous fruits and their derivatives" January 14-16, 2013, funded by Leverhulme Trust, UK.

Roy, S. K. and Khurdiya, D. S. (1982). Keep vegetables fresh in summer. *Indian Horticulture*. 27: 5.

Roy, S. K., Pal, R. K. and Ramanjaneya, K. H. (1999). Jamun: a potential fruit for processing. *Indian Horticulture*. 44(2): 9-11.

Sagrawat, H., Mann, A. and Kharya, M. (2006). "Pharmacological Potential of Eugenia Jambolana: A Review". *Pharmacogenesis Magazice*. 2: 96-104.

Saravanan, G. and Pari, L. (2008). Hypoglycaemic and Antihyperglycaemic Effect of *Syzygium cumini* Bark in Streptozotocin-Induced Diabetic Rats. *Journal of Pharmacology and Toxicology*. 3: 1-10.

Sengupta, P. and Das, P. B. (1965). Terpenoids and related compounds IV, Triterpenoids in the stem bark of *Syzygium cumini* bark. *Indian Journal of Chemical Science*. 42(4): 255-258.

Sepaha, G. and Bose, S. (1956). "Clinical Observations on the Antidiabetic Properties of *Pterocarpus marsupium* and *Eugenia jambolana*". *Journal of the Indian Medical Association*. 27 : 388-391.

Shafi, P. Rosamma, M. and Jamil, K. (2002). "Antibacterial Activity of *Syzygium cumini* and *Syzygium travancoricum* Leaf Essential Oils," *Fitoterapia*. 73: 414- 416. doi:10.1016/S0367-326X(02)00131-4

Shah, R. M., Shah, A. S. and Shah, M. B. (2012). Estimation of ferulic acid in syzygium cumini seeds Extract by HPLC. *International Journal of Pharmaceutical and Chemical Sciences.* 1(1): 201-204.

Shaheer, C. A., Hafeeda, P., Kumar, R., Kathiravan, T., Kumar, D. and Nandanasabapathi, S. (2014). Effect of thermal and thermosonication on anthocyanin stability in Jamun (*Eugenia jambolana*) fruit juice. *International Food Research Journal.* 21 (6): 2189-2194.

Sharma, D. S. (2014). Quality evaluation and storage stability of jamun mango blanded jam. *The Bioscan.* 9 (3): 953-957.

Sharma, P. and Mehta, P. M. (1969). In Dravyaguna vignyan. Part II and Ill, (The Chowkhamba Vidyabhawan, Varanasi, pp. 586.

Shukla, J. P. (1979). Ph. D. thesis submitted to C.S.A. Univ. Agri. And Tech., Kanpur.

Shukla, K. G., Joshi, M. C., Yadav, S. and Bisht, N. S. (1991). Jamun wine making: standardization of a methodology and screening culture. *Journal of Food Science Technology.* 28 (3): 142.

Shyamala G. S. and Vasantha, K. (2010). Phytochemical screening and antibacterial activity of *Syzygium cumini* (L.) (Myrtaceae) leaves extracts. *International Journal of Pharm Tech Research.* 2: 1569-1573.

Singh, A. (1997). Fruit Physiology and Production. Kalyani Publishers, New Delhi. pp. 368-370.

Singh, C. S. and Paswan, V. K. (2015). Process optimization for development of Jamun (*Syzygium cumini* L.) Enriched Shrikhand. *International Journal of Current Microbiology and Applied Sciences.* 4 (12): 73-81.

Singh, I. S. 2001. Minor fruits and their uses. *Indian Journal of Horticulture.* 58: 178 – 182.

Singh, I. S. and Srivastava, A. K. (2000). Genetic diversity Jamun (*Syzigium cumini* Skeels). *Indian Journal of Horticulture.* 45 (3): 2.

Singh, S. K., Krishnamurty, S. and Katyal, S. L. (1967). Fruit culture in India. ICAR, New Delhi: 255-259.

Sood, S. (2015). Studies on the Assessment of Storage Stability of Jamun Jam. *International Journal of Innovative Research and Development.* 4(3): 239-243.

Srivastava, R. P. and Kumar, S. (2003). Unfermented and fermented fruit beverages. *In: Fruit and Vegetable Preservation: Principles and Practices.* International Book Distributing Company, Chaman studio building, 2nd floor, Charbag, Lucknow-226004. pp. 175-193.

Swami, S. B., Thakor, N. J., Haldankar, P. M. and Patil, M. M. (2012a). Processing and value addition in Jamun. *International Journal Processing and Postharvest Technology.* 3 (1): 147-149.

Swami, S. B., Thakor, N. S. J., Patil, M. M. and Haldankar, P. M. (2012). Jamun (*Syzygium cumini* L.) A review of its food and medicinal use. *Food and Nutrition Science.* 3 : 1100-1117.

Thorat, A.V. and Khemnar, M.B. (2014). Development and Sensory Evaluation of Jamun Seed Powder Fortified Cookies. *International Journal of Science and Research.* 4 (10): 184-187.

United States Department of Agriculture. (1912). Bur. of Plant Industry. Inventory of Seeds and Plants Imported by Office of Foreign Seed and Plant Intro. October 1 to December 31st, No. 33. U. S. Dept. of Agric, Washington. 1912.

Vandana, A. K., Suresha, G. J. and Swamy, G. S. K. (2015). Impact of calcium chloride pre-storage treatment on Jamun (*Syzygium cumini* skeels) fruits under cold storage. *The Bioscan.* 10 (1): 199-202.

Veigas, J., Narayan, M. and Laxman, P. (2007). "Chemical Nature, Stability and Bio Efficacies of Anthocyanins from Fruit Peel of *Syzygium cumini* Skeels". *Food Chemistry.* 10: 619-627.

Williamsen, E. M. (2002). Major Herbs in Ayurveda. Churchill Livingstone. 279-282.

Zaman, M. B. and Shariq, M. K. (1995). Hundred drug Plants of West Pakistan. Medicinal plant branch Pakistan Forest Institute Peshawar, 105

Zhang, L. L. and Lin, Y. M. (2009). Antioxidant tannins from *Syzygium cumini* Fruit. *African Journal of Biotechnology.* 8 (10): 2301-2309.

8

Karonda

Scientific Name: *Carissa carandas* Auct.

Family: Apocynaceae

Edible Portion: Epicarp and Mesocarp

India enjoys a prominent position on the pomological map of the world. The varying weather conditions of this country provide suitable environment for growing a variety of fruits. These fruits are available in abundance and also in different seasons. This has resulted in limited scope for expansion of other minor fruits, though they are nutritious, and are the main source of livelihood for the poor (Ravani and Joshi, 2014). Under-exploited fruits (earlier known as minor fruits) have played a very vital role in supplementing the diet of the native people of India (Srivastava *et al.,* 2017). Karonda is a non-traditional fruit crop, which thrives well as a rainfed crop. Once established, the plant hardly needs any care and gives yield with minimum management. *Carissa congesta* Wight (syn. *Carissa carandas* Auct., formerly widely known as *C. carandas* L.), belongs to the dogbane family Apocynaceae (Jain, 1991), found to be widely distributed throughout India. The shrub is commonly known as Karonda (English), Karaunda (Hindi), Karamardaka (Sanskrit), Koromcha (Bengali), Bengal Currant or Christ's thorn (South India), Vakkay (Telugu), Kilaakkaai (Tamil), and Karja Tenga (Assam). It's a hardy, spiny and evergreen indigenous shrub grown widely in India, which bear berry like edible fruits in attractive colour. This fruit crop is monitory reward to tribal communities in various parts of India. *C. carandas* and *C. spinarum* are native to India while *C. grandiflora* is native to South Africa. *C. carandus* is also grown in Sri Lanka, Myanmar, Thailand and Peninsular Malaysia. In India it is found wild in the Western Ghats, Konkan area of Maharashtra and throughout the semi-arid regions. It is widely cultivated in the home gardens, farmer's fields and orchards as hedge plant and occasionally few plants are grown for commercial purpose.

Importance and Uses

Carissa species has been of much socio-economic importance in the tribal area of Gujarat, Maharashtra, Rajasthan, and Madhya Pradesh. Fruit is used for preparation of jelly, pickle and preserve (Singh, 1984). The ripe fruit having antiscorbutic properties is reported to be cooling, acidic and useful in bilious (Watt, 1972). Fruit is slightly sour and astringent in taste, therefore its cultivation is only confined as a fencing bush and it is not popular as desert fruit (Bajpai *et al.,* 2015).

Its fruits are berry-sized, which are commonly used as a condiment or additive to Indian pickles and spices. Karonda is good appetizer, and the fruit is pickled before it gets ripened. Ripe Karonda fruit are sub-acidic to sweet in taste with peculiar aroma and contains high amount of pectin. The fruit may be eaten as a dessert when ripe or used in the preparation of fruit products such as jelly, jam, squash, syrup, tarts, chutney, sauce, *Carissa* cream and jellied salad, which are of great demand in international market (Wani *et al.,* 2013). The unripe fruits are sour and astringent and can be used for pickles, sauce and chutneys. The unripe fruits yield milky white latex, which can also be used in preparing chewing gum and rubber (Monachino, 1946). Fruits can also be used in dyeing and tanning industries and considered as antiscorbutic for cure of anemia. The dried fruits may become a substitute for resins (Cheema and Cheema, 1971). The fruit can be candied just like cherry. Girdhari Lal *et al.* (1967) described the procedure of making Karonda preserve. The Karonda fruit is a rich source of iron and contains a fair amount of Vitamin C. The wine prepared from ripe fruits contains about 14.5 to 15.0 per cent alcohol and is very much liked by wine fanciers (Nalawadi and Jayasheela, 1975).

Karonda bushes are also suitable for hedging in the home gardens, and are sometimes grown as an ornamental plant due to its beautiful cherry-like fruits. The wood of Karonda plant is white, hard and smooth. It used for making spoons and combs. The plant make excellent strong hedge almost impossible to penetrate. The roots of the plant are heavily branched and make it suitable for stabilizing eroding slopes (Noatay, 2004) and its extracts are used in lumbago, chest complaints and veneral disease (Achenbach *et al.,* 1985). Its leaf decoction is used against fever, diarrhea, and ear ache, whereas roots serve as a stomachic, vermifuge, remedy for itches, and insect repellent (Malik *et al.,* 2010).

Traditional Uses

The fruits are traditionally used in the treatments of malaria, epilepsy, nerve disorder, relieve of pain and headache, fever, blood purifier, myopatic spams, dog bite, cough, colds, itches and leprosy (Rahmatullah *et al.,* 2009). The roots are used as anthelmintic, stomachic and antiscorbutic agents and for the treatments of intestinal worm, scabies and pruritus (Warrier *et al.,* 1993). The plant extract of *C. carandas* has been reported to possess cardio-protective, antipyretic and antiviral activities (Taylor *et al.,* 1996; Rajasekaran *et al.,* 1999). The wood of the Karonda plants is white, hard and smooth. It is used for making spoons and combs. The plants can be trained as a strong hedge. The plant is commonly used as a condiment or additive to Indian spices and cold beverages. The sweeter types may be eaten raw but the more acid ones are best stewed with plenty of sugar. Unripe fruit is good appetizer; astringent, antiscorbutic, cooling, acidic, stomachic, anthelmintic and leaf decoctions are given in the commitment of remittent fever (Trivedi and Sharma 2004). Leaf extract is externally applied for curing leprosy. Two drops of plant oil is given with half cup of honey for controlling worms of minors (Trivedi, 2007). Traditional healers of Chhattisgarh use the different plant parts to cover the cancerous wounds and to kill the maggots (Naim, *et al.,* 1988). Traditional healers having expertise in treatment of different types of cancer from *C. carandas.* They use its different plant parts to dress the cancerous wounds and to kill the maggots. To prepare the *C. carandas* decoction, its roots, flowers, spines, leaves and fruits are mixed in equal proportion and crushed to make an aqueous paste. This paste is applied at very initial stages. This paste is boiled in water and when half quantity of water remains, the boiling is stopped and lukewarm decoction is used to wash the cancerous wounds. The healers claim that this decoction is having immense potential to heal the wound and make it infection free. In many ways, it acts in more promising ways than *Azadirachta indica* plant parts. Many healers boil the aqueous paste in mustard seed oil and when all watery contents evaporate, the boiling is stopped and special oil is used for wound dressing.

Composition of Karonda Fruit

Mature fruits contain high amount of pectin and are used for preparation of different products such as jelly, jam, squash, sauce, syrup, *etc.,* which are of great demand in the international market. The unripe fruits are sour and astringent and can be used for pickles and chutney. The dried fruits may become a substitute for raisins (Cheema and Cheema, 1971). The fruits can be candied just like cherry. The wine prepared from ripe fruits contains about 14.5 to 15 per cent alcohol and is very much liked by wine fanciers (Nalawadi and Jayasheela, 1975). The fruits are rich in protein (1.1-2.25 per cent) vitamin C (1.6-17.9 mg/100g) and minerals especially iron (39.1 mg/100g), calcium (21 mg/100g), and phosphorus (38 mg/100g) (Anon., 1950; Anon., 1979; Kumar and Singh, 1993). Fruits sour and astringent in taste are the richest source of iron containing good amount of vitamin C. They are very useful to cure anemia and have antiscorbutic properties also. Daily consumption of 15-20 g Karonda fruits during pregnancy in different forms significantly reduces the prevalence iron deficiency of anaemia in pregnant women (Nandal and Bhardwaj, 2013). Karonda fruit products are highly nutritious, therapeutically important and delicately flavored. If the utilization of Karonda fruits is established in the processing industry, it will give a great fillip to the commercial plantation of this valuable dual-purpose fruit plant (protective hedge as well as nutritive fruits). The composition and food value of ripe and dry Karonda fruits (Anonymous, 1950, 1979) are given in below (Table 1):

Table 1: Proximate Composition of Karonda Fruit

Sl.No.	Constituents	Fresh Fruit (Values/100g edible portion, minerals mg/100g)	Dry Fruit (Values/100g edible portion, minerals mg/100g)
1	Moisture	91	18.2
2	Protein	1.1	2.3
3	Carbohydrate	2.9	67.1
4	Fat	2.9	9.6
5	Fibre	1.5	-

Anonymous, 1950 and 1979; Gehlot ***et al.,*** 2006.

Harvesting and Yield

Karonda bears flower and fruits from the third year onwards and the plants start bearing flowers in the month of February, while fruits are ready by the second week of June upto August. Both unripe and ripe fruits are harvested. Complete harvesting of all fruits may not be done two to three times. Green unripe fruits are available from March to May, which are used for preparation of chutney and pickles, while ripe fruits are harvested in May-June. Colour changes is a good indicator of maturity of fruits *i.e.,* when the colour of the fruits turn to dark purplish black. The spraying of 3 per cent KNO_3 twice after fruit set advanced harvesting and improves quality of Karonda (Sanas, 2014). Fruit yield varies with the soil, climate and management of plants. About 4-5 kg fruit per bush are obtained. Just after harvesting, fruits are kept in shade. Undesirable fruits are stored out and rest packed in shallow baskets cushioned with leaves or straw and marketed. Traditionally, the fresh fruits are collected by tribal people from forest areas and marketed in fresh condition. Pawar (1988) concluded that weight, volume, specific gravity, length, diameter and colour of the fruit can be considered as the physical indices whereas moisture, TSS, sugar, acidity, pH and tannins as chemical indices of maturity of Karonda fruits.

Storage

Karonda has also remained neglected as for as its postharvest handling and technology is concerned. Generally, the tree ripe fruits are harvested. Because of its soft flesh with high moisture content, its shelf life is short. The fruits harvested with stalk intact recorded less shriveling and microbial spoilage (Pawar, 1988). For reduction of spoilage, the cold storage (13 ^{0}C and 94 per cent RH) could be consider the best while for reduction of shriveling, cool chamber could be more effective. Storage life of fruits depends upon the stage of harvest. Fruits harvested at maturity, can be stored for a weak at room temperature, whereas fruits harvested at ripe stage, are highly perishable and last only for 2-3 days. Fruits can be preserved/ stored for 6 months in SO_2 solution (2,000 ppm). Seeds treated with *Trichoderma harzianum* kept in poly bag and stored in refrigerated condition resulted maximum germination 90.00, 80.00, 73.33, 56.66, 50.00 and 43.33 per cent at 15, 30, 45, 60, 75 and 90 days after extraction, respectively. The punnet box packaging remarkably extends the shelf life of Karonda fruits with acceptable good nutritive quality under both ambient and cold storage conditions (Sanas, 2014). The punnet box packed fruits recorded the lowest PLW, lowest percent shriveling and the highest sensory score under both ambient and cold storage conditions. The pre-harvest bagging with butter paper bag was beneficial for quality improvement in Karonda (Sanas, 2014). Seed storage under ambient condition is not recommended as seeds losses its viability early. The use of bio- control agents and different leaf powder are recommended to extend seed viability period and germination in Karonda seeds (Bhavya *et al.*, 2017). The product such as squash, syrup, jam, jelly, pickle, chutney, dried fruits and candy can be successfully prepared from the mature green and ripe fruits. In Southern Florida ripe Karonda fruits are utilized for processing into syrup and wine (Sturrock, 1950). Nalawadi (1967) has reported the preparation of attractive pink coloured jelly from ripe fruits.

Value Added Products of Karonda

Karonda fruits can be processed into number of quality products *viz.*, jam, jelly, chutney, preserve, candy, pickle *etc.*

Jam

Jam is a product made by boiling fruit pulp with sufficient quantity of sugar to a reasonably thick consistency; firm enough to hold the fruit tissues in the position. Jam contains 0.5-0.6 per cent acid and 30-50 per cent invert sugars. According to Fruit Products Order (FPO), a fruit jam should contain minimum 68 per cent of total soluble solids and 45 per cent of fruit portion but, according to Bajpai *et al.* (2015) Karonda pulp 500 g and sugar 850 g with 0.20 per cent acidity was recorded as an ideal recipe for making jam from Karonda fruits. Wani *et al.* (2013) stated that Karonda jam can be stored for at least three months without undergoing any deterioration and evidently (1.0 kg Karonda pulp + 1150 g sugar) showed the best results with regard to physical, chemical and sensory parameters of jam.

Sl.No.	*Ingredients*	*Quantity*	*Sl.No.*	*Ingredients*	*Quantity*
1	Fruit pulp	1 kg	3	Citric acid	1 g
2	Sugar	0.8 kg	4	Water	150 ml

Source: Gehlot *et al.*, 2006.

Technological Flow-Chart for Processing of Jam

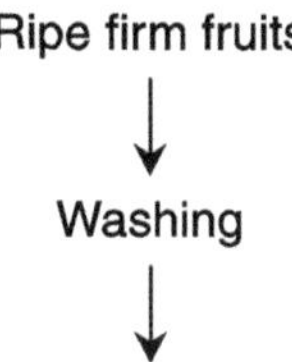

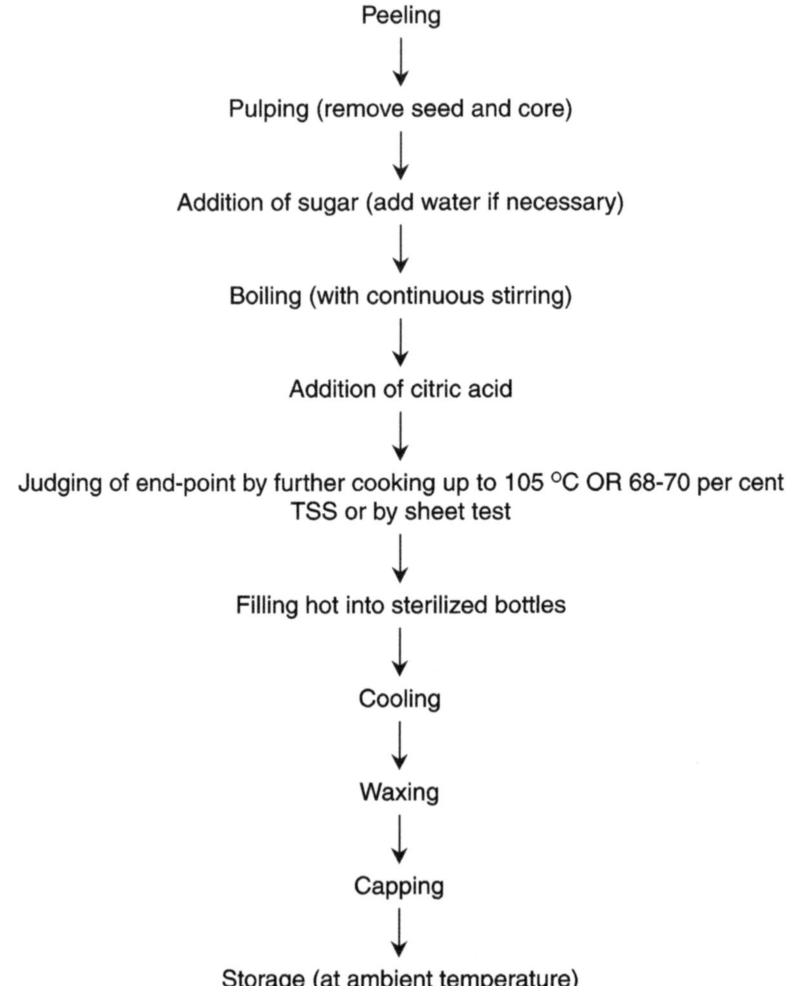

Jelly

A jelly is a semi-solid product prepared by boiling a clear pectin solution of pectin containing fruit extract (free from pulp) after the addition of sugar and acid. A perfect jelly should be transparent, well set, but not too stiff and should have the original flavor of the fruit. It should not be gummy, sticky, syrup or have crystallized sugar in it. As per FPO, a fruit jelly should have minimum 65 per cent of total soluble solids and 45 per cent of fruit portion. As per Bajpai *et al.* (2015) Juice extract and sugar ratio of 1:1.6 with 0.32 per cent acidity was recorded as an ideal recipe for making jelly from Karonda fruits.

Sl.No.	*Ingredients*	*Quantity*	*Sl.No.*	*Ingredients*	*Quantity*
1	Fruit pulp	1 kg	3	Citric acid	1 g
2	Sugar	0.75 kg			

Source: Gehlot *et al.*, 2006.

Technological Flow-Chart for Processing of Jelly

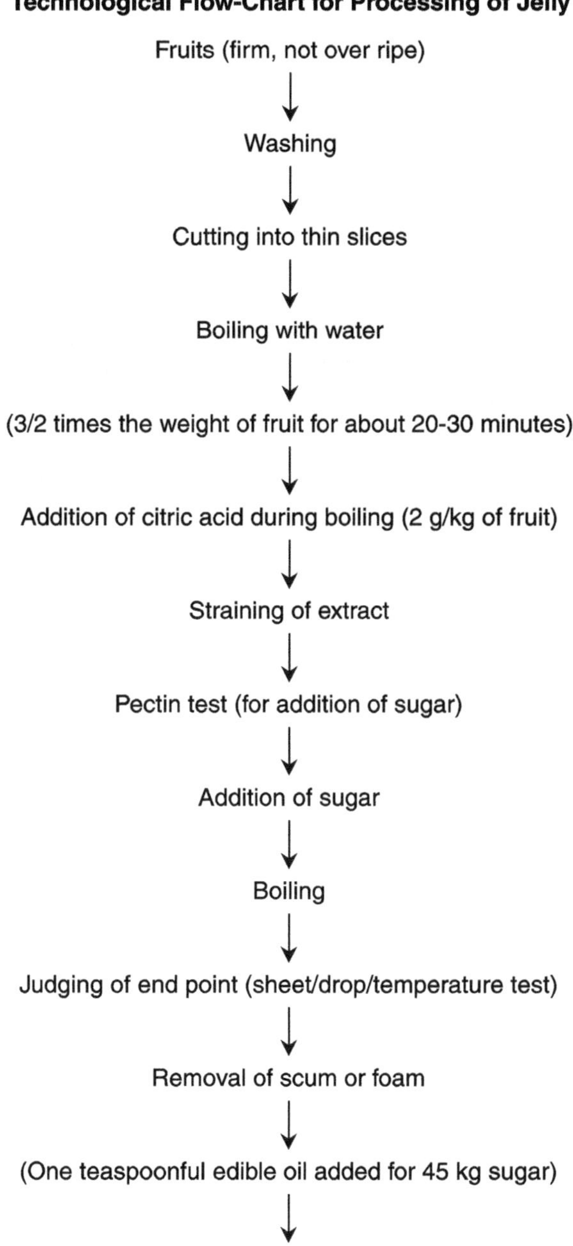

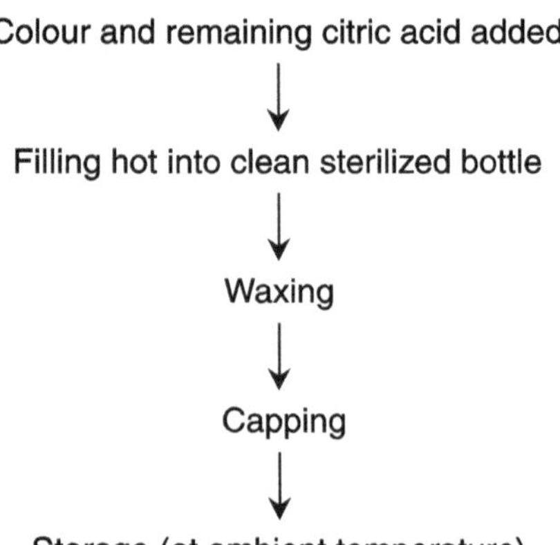

Pickle

Firm and mature fruits were selected, washed and wipe dried. The fruits are individually crushed lightly to creat cracks. Chillies were slit vertically and cut into pieces. For the preparation of cured Karonda pickle the crushed fruits were mixed with salt and allowed to cure for 30 days. After curing all other ingredients (green chillies - 250 g, mustard oil - 300 ml, salt - 250 g, fennel seed - 60 g, mustard seed dhal - 100 g, chilly powder – 10 g, kalunji seeds - 5 g) were mixed thoroughly and stored in a bottle. The contents were stirred on alternate days by shaking the bottles on alternate days of curing (Hiregoudra, 2012). Karonda pickle is easy to prepare and ready to eat. This pickle can be stored for at least four months.

Benefits of Karonda Pickle

- It is a powerhouse of antioxidants.
- It boosts brain power.
- Prevents cancer.
- Reduces cholesterol.

Sl.No.	*Ingredients*	*Quantity*	*Sl.No.*	*Ingredients*	*Quantity*
1	Fruit pulp	1 kg	5	Turmeric	10 g
2	Salt	200 g	6	Fenugreek powder	25 g
3	Onion chopped	50 g	7	Mustard	50 g
4	Cardamom, black pepper, cumin and cinnamon	20 g	8	Cloves (Headless)	4 No.

Source: Gehlot *et al.*, 2006.

Technological Flow-Chart for Processing of Pickle

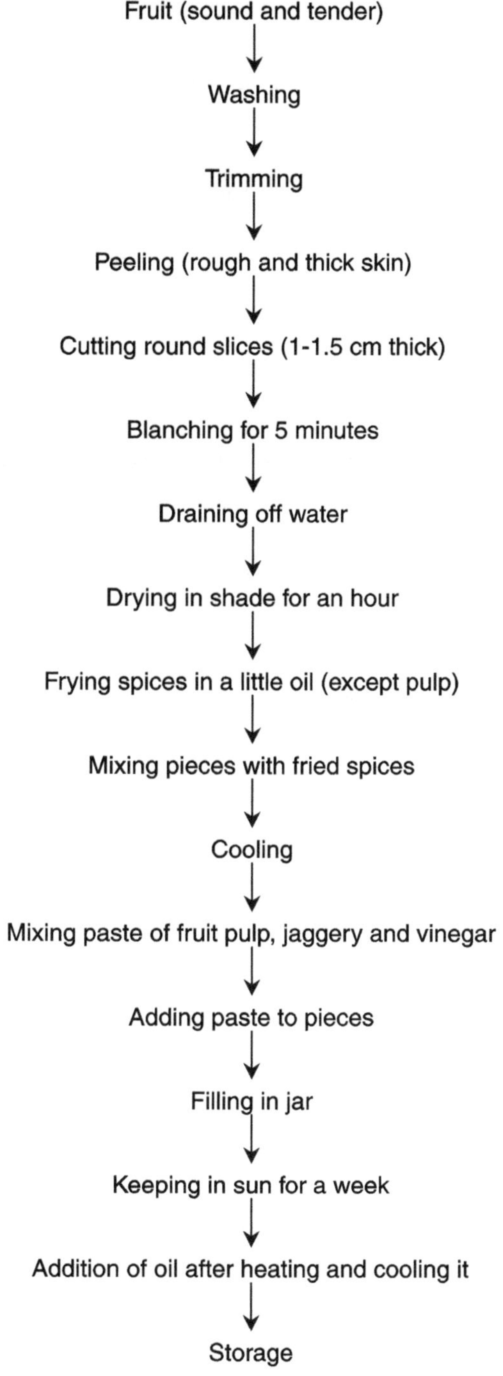

Chutney

Chutney is an important food product prepared both in homes as well as commercially in India. It improves the digestion of food and is also a good appetizer. Karonda chutney is hot, sweet, acidic, spicy, smooth to palate and mellow flavoured. The processing technology of Karonda chutney is similar to that of Karonda jam except that spices, salt and vinegar are also added in processing Karonda chutney.

Sl.No.	*Ingredients*	*Quantity*	*Sl.No.*	*Ingredients*	*Quantity*
1	Fruit pulp	1 kg	5	Ginger chopped	10 g
2	Sugar	0.75 kg	6	Chilli powder	10 g
3	Salt	40 g	7	Hot spices	20 g
4	Garlic chopped	10 g	8	Vinegar	200 ml.

Source: Gehlot *et al.*, 2006.

Technological Flow-Chart for Processing of Chutney

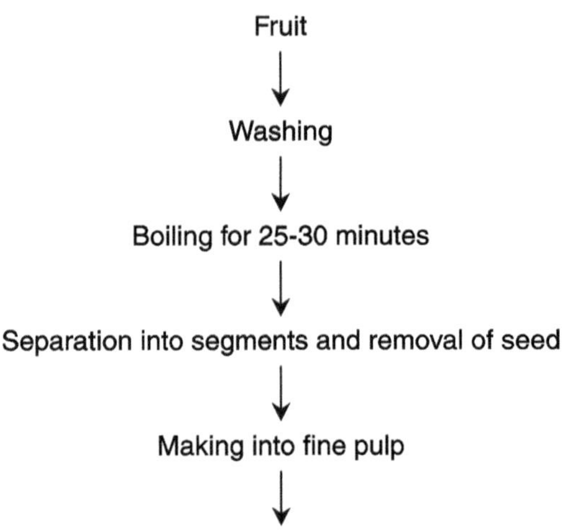

↓

Keeping all ingredients (except vinegar) in cloth bag, tied loosely, putting into mixture and cooking on low flame

↓

During cooking spices bag pressed occasionally

↓

Cooking to consistency of jam (upto105 °C) with occasional stirring

↓

Removal of spice bag after squeezing

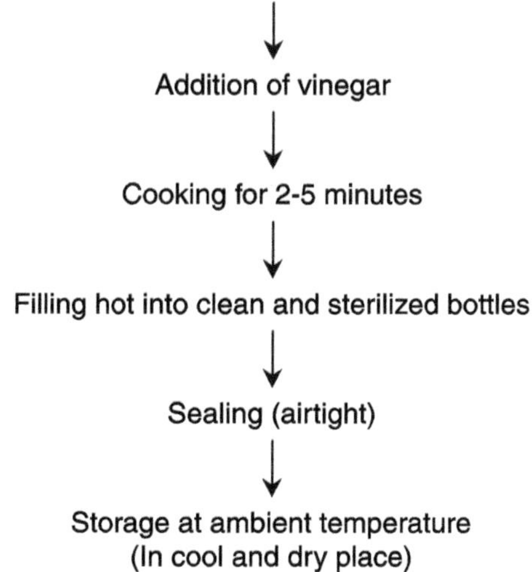

Candy

Karonda fruits impregnated with sugar syrup, subsequently drained free of sugar syrup and dried is known as Karonda candy. The total soluble solids of the impregnated fruits are kept at 75 per cent to prevent fermentation. Invert sugar or glucose (25-30 per cent) may also be substituted for cane sugar to prevent crystallization of sugar. Whitish fruit with pink blush are excellent for making a delicious and nutritious Karonda candy. A study has been conducted by Patil *et al.,* 2014; and Manivasagan, 2004) on Karonda candy in which they prepared the Karonda candy with sugar and citric acid at different ratio and finally dried it by cabinet and solar drier and found that the best results which fruit pieces impregnated with cane sugar (TSS 60 ^{0}Brix)+0.5 per cent citric acid + cabinet drying (Patil *et al.,* 2014a). He also observe the TSS, titratable acidity, reducing sugars and total sugars content of the candy irrespective of drying methods and recipes used in experimentation which was increasing with advancement of storage. Karonda fruits treated with alum 5 per cent prior to osmotic dehydration for three hrs give best result in respect to slight decrease in ascorbic acid, iron content, acidity and increase in moisture content (Suhasini *et al.,* 2015). The microbial growth was below the critical level, *i.e.* $3X10^5$ cfu/g of sample in Karonda candy (Chaudhary *et al.,* 2010).

Sl.No.	*Ingredients*	*Quantity*	*Sl.No.*	*Ingredients*	*Quantity*
1	Fruit pulp	1 kg	3	Citric acid	3 g
2	Sugar	1.5 kg			

Technological Flow-Chart for Processing of Candy

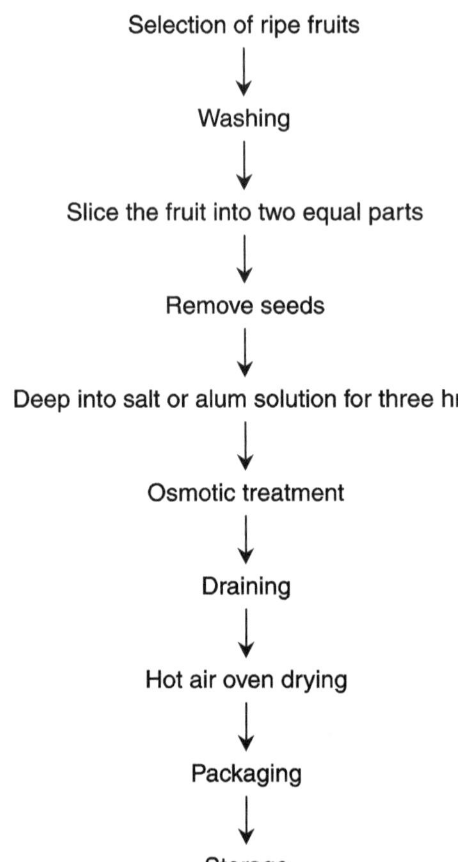

Squash

Bajpai *et al.* (2015) recommended that recipe containing juice and sugar ratio 1:1.5 with the 0.20 per cent acidity was found to be the best followed by recipe containing juice and sugar ratio 1:1.2 with 0.21 per cent acidity.

Technological Flow-Chart for Processing of Squash

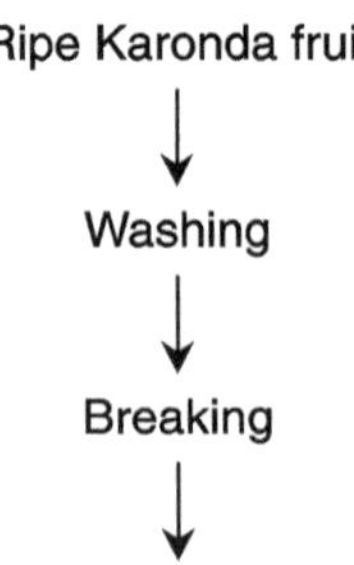

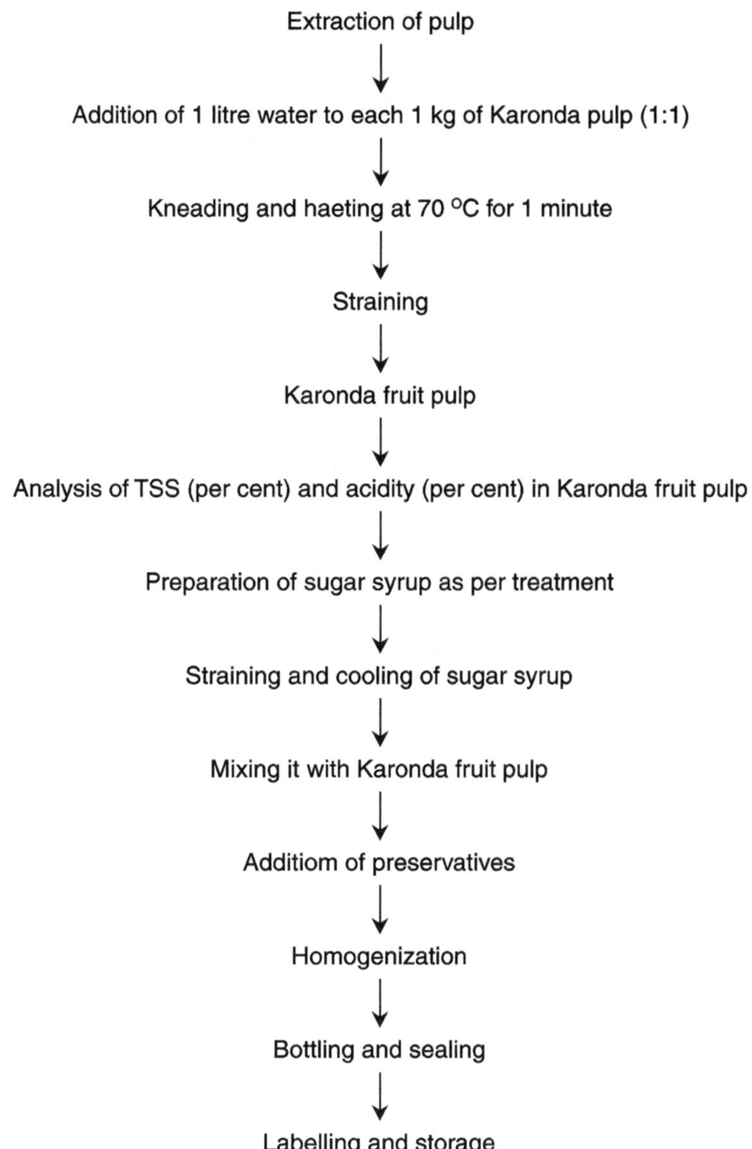

Karonda Powder

Karonda fruits are subjected to blanching at 85 °C for 5 minutes followed by sulphitation with 0.5 per cent KMS for 15 minutes to improve its color. The fruits are then ground coarsely and dehydrated in cabinet drier at 60 ± 1 °C or sun dried. Dried products are then packed and stored in a cool and dry place (Srivastava and Kumar, 2006).

Karonda Beverages

Various types of beverages like ready to serve (RTS) drinks, nectar, squashes can be prepared from Karonda juice/pulp using the methods of Srivatsava and Kumar (2006). Moreover, Karonda juice can be blended with guava, papaya and pineapple juices in different proportions and the combination of Karonda juice with pineapple juice showing best organoleptic quality and acceptability (Shaheel, 2015). The blend of 25 per cent Karonda juice + 75 per cent pineapple juice recorded highest total sugars (10.35 per cent), reducing sugars (6.96 per cent) and organoleptic score (7.42) followed by 50 per cent Karonda juice + 50 per cent guava juice of 7.18 organoleptic value (Shaheel *et al.,* 2015).

Karonda Syrup

The ripe fruits of Karonda are boiled with baking soda and salt. For every cup of juicy pulp half teaspoon of baking soda is added and boiled in one liter of water at 100 ^{0}C. The mixture is then boiled down one half of the original quantity, removing the rising scum in the process and juice is again strained. For every cup, a quarter cup of sugar is added. The mixture is again boiled for 40 minutes. The cooled syrup is poured in to sterilized bottle and sealed (Arif *et al.,* 2016).

REFERENCES

Achenbanch, H., Naibel, R. and Addae, Mer Sah (1989). *Phytochemistry*, 24,: 2325-2328.

Anonymous (1950). In: The wealth of India: A Dictionary of Indian Raw Materials and Industrial Products. Vol. 3, CSIA, New Delhi: 22.

Anonymous (1979). Extension Bullentin, IIHR, Bangalore, No. II: 34-35.

Arif, M., Mehnaz K., Jawaid, T., Khalid, M., Saini, K. S., Kumar, A. and Ahmad, M. (2016). *Carissa carandas* Linn. (Karonda): An exotic minor plant fruit with immense value in nutraceutical and pharmaceutical industries. *Asian Journal of Biomedical and Pharmaceutical Sciences.* 6(58): 14-19.

Bajpai, R., Yadav, M., Mure, S. and Kushwah, R. S. (2015). Browning analysis of different Karonda processed products during storage. *Plant Achieve.* 15(1): 339-342.

Bhavya, N., Naik, N., Masuthi, D. K. and Sabarad, A. (2017). Studies on effect of different storage conditions on viability of Karonda seeds. *International Journal of Current Microbiology and Applied Sciences.* 6(9): 1057-1066.

Chaudhary, R., Yadav, M. and Singh, D. B. (2010). Microbial analysis of different Karonda processed products during storage. *Indian Journal of Horticulture.* 67(1): 144-146.

Cheema, G. S. and Cheema, B. K. (1971). *Farm Journal.* Calcutta, 12:24.

Gehlot, R., Dhawan, S. S. and Singh, K. (2006). Processed products of Karonda. *Food and Pack.* 6(6): 28-30.

Girdhari, Lal., Siddappa, G. S. and Tandon, G. L. (1967). In: Presrvation of Fruits and Vegetables, ICAR, New Delhi: 182.

Hiregoudra, V. S. (2012). Physico-chemical characteristics, value addition and shelf life of evaluation Karonda (*Carissa carandas*). M. Sc. Thesis UAS, Dharwad.

Jain, S. K. (1991). Dictionary of Indian Folk Medicine and Ethnobotany. New Delhi: Deep Publications.

Kumar, S. and Singh, I. S. (1993). Variation in quality traits of Karonda (*Carissa carandas* L.) germplasms. *South Indian Horticulture*. 41(2): 108-109.

Malik, S. K., Chaudhury, R., Dhariwal, O. P. and Bhandari, D. C. (2010). Genetic Resources of Tropical Underutilized Fruits in India. NBPGR, New Delhi.: 168.

Manivasagan, S. (2004). Evaluation of Karonda (*Carissa cardandas* Linn) cultivars for processing, CCS Haryana Agricultural University, Hisar, India, M.Sc. Thesis 2004.

Monachino, J. (1946). A resume of the American Carisseae (Apocynaceae). Lloydia. 9(4): 293-309.

Naim, Z., Khan, M. A. and Nizami, S. S. (1988). "Isolation of a new isomer of ursolic acid from fruits and leaves of *Carissa carandas*". *Pakistan Journal of Scientific and Industrial Research*. 31 (11): 753-55.

Nalawadi, U. G. (1967). Food Industries Journal. December-January, 1967.

Nalawadi, U. G. and Jayasheela, N. (1975). *Progressive Horticulture*. 7: 37-38.

Nandal, U. and Bhardwaj, R. L. (2013). Role of fruit juices in nutritional and health security. *Indian Food Packer*. 67(2): 113-122.

Noatay, K. L. (2004). Karonda the pickle berry, *India Tribune*, http://www.tribuneindia.com/2004/20040301/agro.htm#5

Patil, Rashmi, Raut, V. U. and Wankhade, R. S. (2014). Studies on preparation of Karonda candy. *International Journal of Processing and Postharvest Technology*. 5 (2) : 136-140.

Patil, Rashmi, Raut, V. U. and Wankhade, R. S. (2014a). Sensory quality and economics of preparation of Karonda candy. *International Journal of Processing and Postharvest Technology*. 5 (2) : 169-172.

Pawar, C. D. (1988). Studies of postharvest handling and preparation of different products of Karonda (Carissa carandas L.) fruits, M.Sc. (Agri.) Thesis, Dr. Balasaheb Sawant Konkan Krishi Vidyapeeth, Ratnagiri (India).

Rahmatullah, M., Noman, A., Hossan, M. S., Harun Or-Rashid, M., Rahman, T., Chowdhury, M. H., and Jahan, R. (2009). A survey of Medicinal Plants in Two Areas of Dinajpur District, Bangladesh Including plants which can be used as Functional Foods. *American-Eurasian J. of Sust. Agril.*, 3(4), 862-876.

Rajasekaran, A., Jeyasudha, V., Kalpana, B., and Jayakar, B. (1999). Preliminary Phytochemical and Antipyretic Evaluation of *Carissa carandas*. *Indian Journal of Natural Product*. 15(1), 27-29.

Ravani, A. and Joshi, D. C. (2014). Processing for value addition of underutilized fruit crops. *Trends in Postharvest Technology*. 2(2): 15-21.

Sanas, M. P. (2014). Effect of post flowering foliar nutrient sprays and bagging on Karonda and postharvest management in Karonda (*Carrissa conjesta* Linn.). Ph. D. (Hort.) thesis at Dr. BSKKV, Dapoli, Ratnagiri, Maharastra.

Shaheel, S. (2015). Studies on the value addition of Karonda (*Carissa carandas* L.) juice by blending with guava, papaya and pineapple juices for rts beverage. M. Sc. Thesis at Dr.YSR University Andhta Pradesh.

Shaheel, S. K., Swami, D. V., Kumar, B. P. and Krishna, K. U. (2015). Effect of blending of Karonda (*Carissa carandas* L.) juice with guava, papaya and pineapple juices on its quality and organoleptic evaluation. *Plant Archives*. 15(1): 187-192.

Singh, I. S. (1984). Fruit vegetable utilization and preservation (in Hindi). Directorate of Extension, N. D. Univ. of Agric. and Tech., Faizabad.

Srivastava, A., Sarkar, P. K. and Bishnoi, P. K. (2017). Value addition in under-exploited fruits of Karonda (*Carissa carandus* L.): an earning opportunity for rural communities in India. *Rashtriya Krishi*. 12 (2): 161-163.

Srivatsava, R. P. and Kumar, S. (2006). Principles of fruits and vegetables. 3rd revised and enlarged edition. 235-238 pp.

Sturrock, D. (1950). In: fruits of Southern Florida. A Hand Book of Home Owner. Horticultural Books, Surat, Florida.: 73.

Suhasini, L., Vanajalatha, K., Padmavathamma, A. S. and Rao, P. Vekateshwar. (2015). Physico-chemical properties of osmotically dehydrated Karonda (*Carissa carandas* L.). *Asian Journal of Horticulture*. 10(1): 53-59.

Taylor, R. S. L., Hudson, J. B., Manandhar, N. P., and Tower, G. H. N. (1996). Antiviral Activities of Medicinal Plant of Southern Nepal. *Journal of Ethno-pharmacology*, 53(2), 97-104.

Trivedi, P. C. (2007). "Ethanomedicinal plants of India," Aavishkar Publishers, Churna Rasta Jaipur (Raj.) India: 71.

Trivedi, P. C. and Sharma, N. K. (2004). "Ethanomedicinal plants." Pointer publisher, Jaipur (Raj.) India: 38.

Wani, R. A., Prasad, V. M., Hakeem, S. A., Sheema, S., Angchuk, S. and Dixit, S. (2013). Shelf life of Karonda (*Carissa carandas* L.) jams under ambient temperature. *African Journal of Agricultural Research*. 8(21): 2447-2449.

Warrier, P. K., Nambiar, V. P. K., and Ramankutty, C. (1993). Indian Medicinal Plants: A Compendium of 500 Species (Vol. I). Universities Press (India) Pvt. Ltd.

Watt, G. (1972). Dictionary of the economic products of India. Periodical experts, Delhi: 165-166.

9

Phalsa

Scientific Name: *Grewia subinaequalis* L.

Family: Moraceae

Edible Portion: Mesocarp and Epicarp

Phalsa (*Grewia subinaequalis* L.) is a minor fruit of Indian origin. Phalsa is known by different name in different Indian languages such as Phalsa, Dhamin, Parusha, and Shukri in Hindi, Dhaman in Punjabi, Man-Bijal in Assamese, Phalsa and Shukri in Begali, Mirgi Chara and Pharasakoli in Oriya, Phalsa in Gujrati, Phalsi in Maharashtra, Jana, Nallajana, Phutiki in Telagu, Palisa, Tadachi in Tamil, Buttiyudippe and Tadasala in Kannada (Tripathi, 2009; Singh and Singh, 2018). The genus *Grewia* belongs to the family tiliaceae comprising approximately 150 species which include small trees and shrubs, distributed in subtropical and tropical regions of the world among which nearly 40 species are found in India some of which are very for well known their medicinal properties. Most of the genus of Tiliaceae family are wild and known for their fodder, fuel wood, craft works, timbers and therapeutic values *viz. Grewia flavescens* A. Juss, *Grewia villos* and *Grewia hirsuta*. Grewia is the only genus in family Tiliaceae with edible fruits. Extensively cultivated species for their fruit values are *G. subinaequalis* DC. (syn.*G. asiatica.*) and *Grewia tenax* (Frosk.). Medicinal values of Grewia species is due to the presence of different metabolites like saponins, coumarins and anthraquinone. It is a minor fruit and is being cultivated on very small scale in each state. However, in Punjab, Haryana and Uttar Pradesh it is cultivated near cities commercially. It is native to the Indian subcontinent and Southeast Asia (Maury *et al.,* 2012).

The fruit's, leaves and barks of *Grewia* species have high medicinal values and are widely used for the treatment of various common diseases. The fruit of Phalsa are small, 1.0 to 1.9 cm in diameter, 0.8 to 1.6 cm in vertical height, 0.5 to

2.2 g in weight and bears in clusters of 2-8 on leaf axils. When ripe, the fruit skin colour turns from light green to cherry red or purplish red and finally becoming dark purple or nearly black. The fruit is covered with a very thin, whitish blush and becomes soft and tender. The delicate, fibrous flesh is greenish-white in colur, becoming purplish-red during ripening.

Nutritional Importance and Bioactive Compounds

Ripe fruits of Phalsa contain high amount of vitamin (A, C), minerals (calcium, phosphorous and iron) and fiber however, low in calorie and fat. Unripe fruits reported to alleviate inflammation and are administered in respiratory, cardiac, and blood disorders, as well as in fever reduction (Morton, 1987). While, Ripe fruits of Phalsa are consumed fresh, as desserts, or processed into refreshing fruit and soft drinks enjoyed in India during hot summer months as it has cooling tonic and aphrodisiac effects which overcomes thirst and sensation. The bark is used as a soap substitute in Burma. A mucilaginous extract of the bark is useful in clarifying sugar (Singh ans Singh, 2018).

Phytochemical screening revealed the presence of alkaloids, carbohydrates, glycosides, proteins and amino acids, saponins, steroids, acids, mucilage, fixed oils and fats. The pulp contains higher concentrations of phosphoserine as compared to other free amino acids, while the hydrolyzed product contained aspartic acid, glycine, and tyrosine in large amount. Fruit pulp and seeds contain essential amino acids like threonine and methionine, respectively. Whereas, phosphoserine, serine and taurine are the dominant amino acids present in fruit juice. The attractive crimson red to dark purple colour of Phalsa fruit is due to anthocyanin pigments mainly, delphinidin-3-glucoside, cyanidin-3-glucoside and pelargonidin-3, 5-diglucoside. The major phytochemical compounds present in the fruit of Phalsa are triterpenoids, fatty component, flavonoids (quercetin, quercetin-3-O-β-D-glucoside and naringenin-7-O-β-D-glucoside), steroids, saponins and tannins. The fruit posses' very high antioxidant activity due to presence of vitamin C, phenolics, flavonoids, tannins and anthocyanins. In the fruit, highest antioxidant activities are found in fruit peel followed by pulp and seeds. The air-dried Phalsa seeds are also rich in linoleic acid, besides containing fair amount of palmitoleic, heptadecanoic, linolenic and arachidic acids.The nutritive composition of Phalsa fruit is given below.

Nutritional Value of Phalsa Fruit (per 100 g of edible portion)

Nutrients	*Nutrient Values/100 g Fruit*
Calories (Kcal)	90.5
Calories from fat (Kcal)	0.0
Moisture (per cent)	76.3
Fat (g)	<0.1
Protein (g)	1.57

Nutrients	*Nutrient Values/100 g Fruit*
Carbohydrates (g)	21.1
Dietary Fiber (g)	5.53
Ash) (g)	1.1
Calcium (mg)	136
Phosphorus (mg)	24.2
Iron (mg)	1.08
Potassium (mg)	372
Sodium (mg)	17.3
Vitamin A (μg)	16.11
Vitamin B_1, Thiamine (mg)	0.02
Vitamin B_2, Riboflavin (mg)	0.264
Vitamin B_3, Niacin (mg)	0.825
Vitamin C, Ascorbic acid (mg)	4.385

Source: Yadav (1999).

Medicinal Value

Phalsa has a very high value in the indigenous system of medicine. Almost all the species and whole plant part of the genus Grewia is utilized as a folk medicine in different diseases and disorders. *Grewia asiatica* is a food plant and can also be used as a herbal medicine for the treatment of various diseases such as cancer, ageing, fever, rheumatism and diabetes. This plant can also be used as an antioxidant and radio protective agent. Its fruits possess a variable extent of antioxidant activity besides providing essential nutrients. The fruit of *G. asiatica* (Syn. *G. subinaequalis*) has astringent and cooling effect. The fruits are beneficial for heart and liver disorders, anorexia, indigestion, thirst, toxemia, stomatitis, hiccough, asthma, spermatorrhoea and diarrhoea and are used for treating throat, tuberculosis and sexual debility troubles. The fruit also has radio-protective effect. Aqueous extracts of leaves and fruits posses anti-cancerous property against liver and breast cancer. Thus, fruits and leaves extracts of Phalsa can be utilized for the management of human cancer. Fruit is also beneficial against throat ailments.

The bark is used as a soap substitute in Burma. A mucilaginous extract of the bark is useful in clarifying sugar. An infusion of the bark is given as a demulcent, febrifuge, and treatment for diarrhoea. The leaves are believed to have antibiotic properties hence, applied on skin eruptions and they are known to have antibiotic action (Singh and Singh, 2018). Its juice may be taken to control the diabetes due to a low glycemic index. Carbohydrates break down more slowly in low glycemic index foods. Low glycemic index foods are also believed to reduce the risk of coronary heart disease and obesity.

Harvesting and Yield

For uniform ripening of Phalsa fruits, apices of shoots may be pinched in mid-May to check further shoot growth. Fruits start ripening in the first week of June and continue for a month. Only a few fruits in a cluster ripen at any one time, so continuous harvesting is necessary or harvested twice a week. Average yield per plant is 9-11 kg in a season. After harvesting fruits are packed in small baskets of 2 kg capacity or in packs. Phalsa fruit is perishable in nature, hence should be transported to the market soon after harvesting.

Storage

Being highly perishable in nature, the fruit must be utilized within 24 hours after picking. Fruits when harvested at unripe mature stage, can be stored up to 48 hours while, fully ripe fruit cannot be stored more than 24 hours at ambient conditions. Fruits harvested at turning stage can be stored about a week at 7 °C in cold storage (Tiwari *et al.*, 2014). According to Ray and Bala (2016) stated that the Phalsa fruits of tall cultivar can be stored for 10 days in deep-freeze, 8 days in refrigerator and only a single day or 24 hours at ambient temperature as compared to dwarf which can be stored for 9 days in deep-freeze, 7 days in refrigerator and only a single day at ambient temperature on the basis of per cent weight loss and visual characteristics. The popularity of Phalsa fruit is due to its attractive colour ranging from crimson red to dark purple and its pleasing taste. The juice when extracted gives a deep crimson red to dark purple colour and is very popular. It is rated very high in indigenous system of medicine. The juice is extremely refreshing and is considered to have a cooling effect especially in hot summer. Heating the crushed Phalsa fruit to 50 °C gives the highest recovery of the juice with an appropriate quantity of anthayanin and other soluble and insoluble materials. Studies have shown that addition of cane sugar to the juice has a protective effect on colour stability.

Processing

Postharvest losses in Phalsa are very high and it can be managed by value addition in fruits. Phalsa juice is very popular due to its pleasing flavor and deep crimson-red colour. In addition to extremely refreshing quality of Phalsa juice, it can be processed into ready-to-serve (RTS) and carbonated beverages. Syrup and squash can also be prepared with Phalsa fruit juice after mixing with sugar and preserved with sodium benzoate. The fruits are highly perishable in nature and due to its perishability it cannot be exported but its processed products are very appreciable. Ripe fruits are consumed fresh in desserts, or processed into refreshing soft drinks like squash, RTS, Sherbet *etc.* which are enjoyed during hot summer months in India.

Value Added Products of Phalsa

1) Phalsa Juice

Flow-Chart for Juice Extraction (Tiwari *et al.,* 2014)

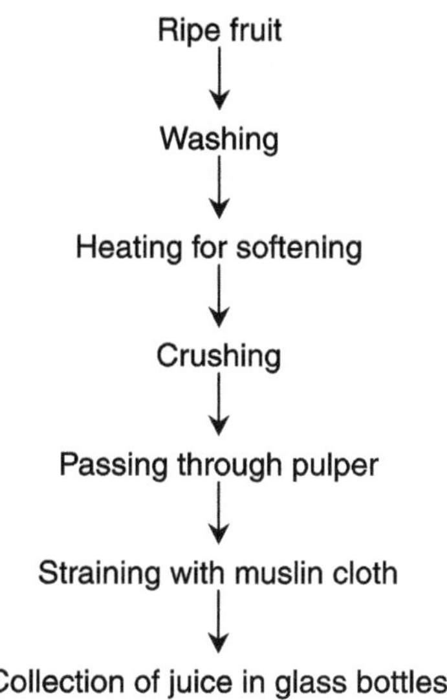

2) Phalsa RTS

Flow-Sheet for Preparation of Phalsa RTS (Tiwari *et al.,* 2014)

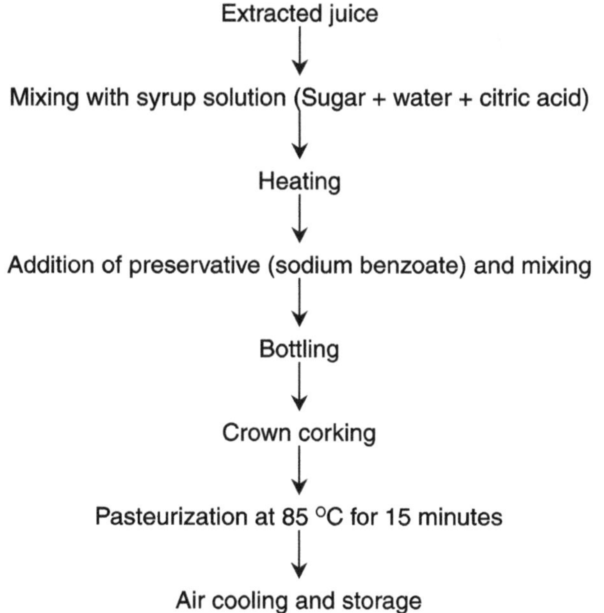

3) Phalsa Squash

Flow Sheet for Preparation of Phalsa Squash (Tiwari *et al.*, 2014)

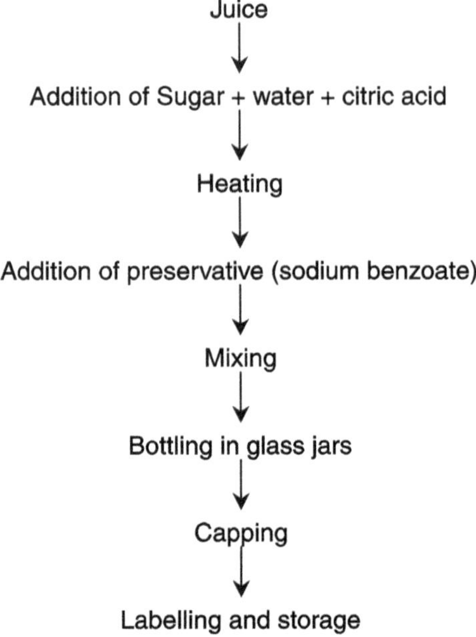

4) Phalsa Sherbet

Flow sheet for preparation of Phalsa Sherbet (Tiwari *et al.*, 2014)

Soaking of fresh Phalsa fruits in water for 30 minutes

↓

Strain out the fruits

↓

Mash the fruit and remove the seeds

↓

Add ground cumin, rock salt and jaggery or sugar to the water

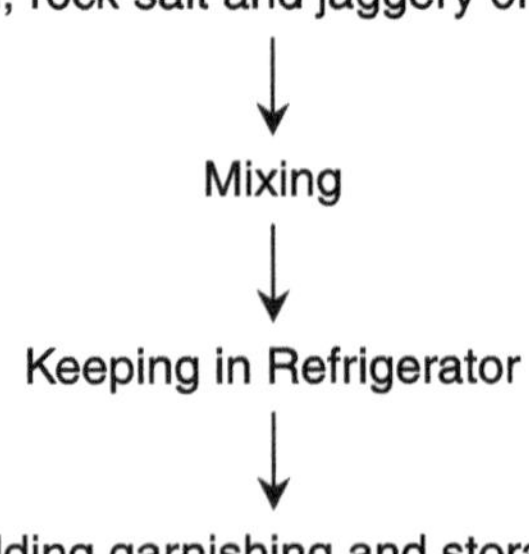

5) Phalsa and Pear Blend Crush

Pangotra *et al.* (2018) standardized Phalsa and Pear crush by adding desired quantity of sugar and citric acid in warm water and the solution was strained off through muslin cloth. The solution was then added in Phalsa-pear blend so as to maintain its total soluble solids as 55 °Brix and an acidity of 1 per cent. The crush prepared was filled in pre-sterilized glass bottle crown corked, processed for 30 min. in boiling water, cooled immediately, labeled. In the freshly prepared blended crush, highest reducing sugar (32.62 per cent), total sugar (43.51 per cent), ascorbic acid (7.38 mg/100 ml) and anthocyanin (18.92 mg/100 ml), tannin (0.43 mg/100ml), iron (0.31 mg/100 ml) and phosphorus (6.87 mg/100 ml) were recorded in treatment T1 (100:00::Phalsa: pear) whereas the lowest value were recorded in treatment T11 (50:50::Phalsa: pear). Sensory evaluation of blended crush revealed that the highest score of colour (7.72), body (7.34), aroma (7.45), taste (7.44) and overall acceptability (7.48) was recorded in treatment T5 (80:20 Phalsa: pear), respectively.

REFERENCES

Maury, P. K., Jain, S. K. and Lal, N. (2012). *Grewia asiatica*: An overview. *International Journal of Pharmaceutical Research and Development.* 4 (04): 154-158.

Morton, J. F. (1987). Phalsa. *In*: Fruits of warm climates. Julia Morton, Miami, FL, pp. 276-277.

Ray, A. and Bala, K. (2016). Effect of storage condition on sensory attributes of Phalsa fruit (*Grevia asiatica*) of *var*. Sharbati. *International Journal of Advance Research in Science and Engineering*. 5(8): 720-725.

Singh, K. K. and Singh, S. P. (2018). Cultivation and utilization in Phalsa (Grewia asiatica L.) under Garhwal Himalayas region. *Journal of Medicianal Plant Studies.* 6(1):254-256.

Sinha, J., Purwar, S., Chuhan, S. K and Rai, G. (2015). Nutritional and medicinal potential of *Grewia subinaequalis* DC. (syn. *G. asiatica.*) (Phalsa). *Journal of Medicinal Plants.* 9(19): 594-612.

Tiwari, D. K., Singh, D., Barman, K and Patel, V. B. (2014). Bioactive Compounds and Processed Products of Phalsa (*Grewia subinaequalis* l.) Fruit. *Popular Kheti.* 2(4): 128-132.

Tripathi, P. C. (2009). Phalsa (*Gravia subinaequalis*) cultivation. National Research Centre for Women in Agriculture Opposite Kalinga studio P.O. Bara munda, Bhubaneswar-751 003 (Orissa).

Yadav, A. K. (1999). Phalsa: A Potential New Small Fruit for Georgia. p. 348–352. In: J. Janick (ed.), Perspectives on new crops and new uses. ASHS Press, Alexandria, VA.

Zia-Ul-Haq, M., Stankoviæ, M.S., Rizwan, K and Feo, V. D. (2013). Review *Grewia asiatica* L., a Food Plant with Multiple Uses. *Molecules.* 18: 2663-2682.

10

Mahua

Scientific Name: *Madhuca longifolia*

Family: Sapotaceae

Mahua, the Indian Butter Tree (*Madhuca longifolia* (Koenig) J.F. Macribide) is an important tree having vital socioeconomic value and grow throughout the tropical and subtropical region of the Indian subcontinent (Dhakar *et al.,* 2015). Its known as Indian butter tree (in English), Mahua, Mohwa, Mauwa (in Hindi) Mahwa, Maul, Mahwla (in Bangali), Mahwa, Mohwra (in Marathi), Madhuda (in Gujrati), Ippa (in Telugu), Illupei, Ewpa (in Tamil), Tuppe (in Kannad), Poonam, Ilupa (in Malyalam), Mahula, Moha, Madgn (in Oriya) by Bina *et al.,* 2010. It is a deciduous tree that grows widely under dry tropical and subtropical climatic conditions. It is very hardy and thrives well on rocky, gravely, saline and sodic soils, even in pockets of soil between crevices of barren rock (Singh, 1998). It is one of the multipurpose forest tree species that provide an answer for the three major Fs *i.e.* food, fodder and fuel (Patel *et al.,* 2011). Mahua is a tree valued for its fruit, seeds, which are the largest source of natural hard fat commercially known as Mahua butter or mowrah butter. Fruits are eaten as raw or cooked. The fruit pulp may be utilized as source of sugar, whereas the dry husk makes a good source of alcoholic fermentation. Seeds are good source of oil (Singh *et al.,* 2005).

A short postharvest longevity remains a major limiting factor for many plants and their products. Mahua (*Madhuca indica* Syn. *Bassia latifolia*) is one plant, belonging to the family Sapotaceae, which has great economic value. It grows naturally in many parts of the world and its sugar rich edible corollas have great nutrient value. Indeed, the Mahua is used as food in several countries, especially India (Anonymous, 1995). Tribal people comprise nearly 23 per cent of the total population of the state of Orissa, India. Driven by hunger, they use this plant product as their staple source of food during most parts of the year. The flowers have an antibacterial activity against *Escherichia coli* (Sujatha and Das, 1988) and are effective against rice pest diseases (Verma and Singh, 1979). The plant is also highly efficacious in eradicating many other diseases (Singh and Agrawal, 1989; Wayland, 2004). The brew extracted from the flowers, called "Mahuli" in the local dialect, is a part of the lives, tradition and culture of the tribal people. It has been estimated that 90 gallons of 95 per cent alcohol can be extracted from 1 ton of Mahua flowers (Anderson, 2000). Thus, it plays a very important role in tribal economy (Rout and Das 1993; Varghese *et al.,* 2002). India is a land with diverse group of people. The country is a home for more than 400 tribal groups, who still depend upon traditional agriculture and forest for their livelihood and survival (Mishra *et al.,* 2001).

Nutritional Properties of Mahua flower

Constituents	*Value*
Moisture (per cent)	19.8
Protein (per cent)	6.37
Fat (per cent)	0.50
Reducing Sugar (per cent)	50.62
Total Inverts (per cent)	54.24
Cane Sugar (per cent)	3.43
Total Sugar (per cent)	54.06
Ash (per cent)	4.36
Calcium (per cent)	8.00
Phosphorus (per cent)	2.00

Source: Kureel *et al.*, 2009.

Variability in Oil Content, Fatty Acid Profile and Biodiesel Traits of Mahua

Parameter	*Range*	*Mean*
Kernel oil (per cent)	44.43-61.50	54.37
Palmitic oil (per cent)	11.65-25.86	19.56
Stearic oil (per cent)	19.11-32.16	24.78
Oleic oil (per cent)	32.91-48.65	42.87
Linoleic oil (per cent)	9.36-15.42	12.64
O/L ratio	2.39-4.48	3.43
Total saturated fatty acid (per cent)	39.33-52.75	44.34
Total unsaturated fatty acid (per cent)	46.66-60.65	55.52
Saturation ratio	0.64-1.13	0.80
Saponification value (SV)	198.34-202.78	201.76
Iodine value (IV)	52.09-68.59	61.46
Cetane number (CN)	58.00-61.58	59.52

Source: Yadav *et al.* (2011).

Uses

Madhuca indica has several pharmacological activity, and potential to provide health to the society. It is used as Anti diabetic, antiulcer, hepato protective, anti pyretic, anti fertility, analgesic, anti oxidant, swelling, inflammation, piles, emetic, dermatological, laxative, tonic, anti burn, anti earth worm, wound healing headache and many more problems (Patel *et al.,* 2012). Every part of any plant posses some medicinal properties, either in small of large proportion, meanwhile some parts of a plant often contain a quit different active ingredients, so that one part may be toxic and another one quite harmless (Erik and Michael, 2004). Mahua is the most valuable tree because every part of it is put into use such as food, fodder and fuel. Several parts of the tree, including the bark, are used for their medicinal properties. It is considered holy by many tribal communities because of its usefulness (Owoabi *et al.,* 2007). Detoxified Mahua seed flour appears to be a good source of protein for food and feed products (Singh and Singh, 1991). The plant consist of several parts, they may be classified according to the function. They are root, bark, leaves, flowers, fruits, seeds, oil *etc.*

Part of the Plant	*Medicinal Properties*
Leaf oil	Enzyma, Wound Healing, Anti Burns, Bone Fracture
	Emollient, Skin Disease, Rheumatism, Headache, laxative, Piles, Hemorrhoids, Emetics, Anti Earth worm.
Fruit	Sweet, Refrigerant, Aphrodisic, Tonic, Dipsica, Bronchitis, Astringent, Anti Ulcer, Acute and Chronic Tonsillitis, Pharyngitis
Bark	Rheumatism, Ulcer, Inflammation, Bleeding, Spongy Gums, Tonsillitis, Diabetic, Stomach Ache, Anti Snake Poisoning, Astringent, Emollient, Fracture, Itching
Flower	Refrigerant, Liquor, Jelly, Sweet Syrup, Expectorant, Increase the production of milk in woman, Stimulant, Diuretics, Anthelmentic, Strangury, Verminosis, Hepatoprotective. Gastropathy

Source: The wealth of India, 2007; Seshagiri and Gaaikawad 2007, Ingle and Dhabe, 2011.

Sri Laxmidhar Mohanta is the farmer of Keonjhar district of Odisha Developed yellow sticky pot trap by using Mahua oil. He develop this technology by using locally available material *i.e.* earthen pot and Mahua oil (*Madhuca indica*). The outer part of the earthen pot was painted with enamel yellow paint and smeared with Mahua oil. The pot was placed with wooden tag in the field @ 20 numbers per ha. The colour attracts the insects and get stuck to the pot due to stickiness of Mahua oil. The performance of yellow sticky pot trap is at par with the trapping efficiency of white fly to commercial yellow sticky traps. It is cheaper, easy to prepare and eco-friendly; which controls significantly the viral diseases like little leaf in brinjal, leaf curl in tomato, YVMV in okra and mosaic in cucurbits, Further, it solves the problem of unavailability of commercial trap in the market (IHC, 2014).

Traditional Uses

In spite of being a rich source of nutrition and easy availability in the rural areas these flowers are not very popular as food. Only a small quantity of flowers is consumed raw, cooked or fried in different parts of India (The wealth of India, 1962). Flowers are used as tonic, analgesic and diuretic. Flowers have been traditionally used as cooling agent, tonic, aphrodisiac, and astringent, demulcent and for the treatment of helminthes, acute and chronic tonsillitis, pharyngitis as well as bronchitis (Chandra, 2011). Also used as anthelmintic, demulcent, laxative, stimulant and tonic, fomentation with dried ones produces relief in orchids', decoction used as expectorant; beneficial in impotence due to general debility when administered with milk (Sinha *et al.,* 2017). Major quantity of flowers is used in the preparation of distilled liquors. The freshly prepared liquor has a strong, smoky foetid odour, which disappears on ageing. Apart from food and liquor, the flowers are also used as a source of cattle feed (The wealth of India, 1962). The wood of Mahua tree is also used in the household utility like door and window making (Kirtikar and Basu, 2001).

The Mahua tree is having lots of nutritional value in it. It produces fruit, which is valued for its seed which yield high quantity of fat commercially known as Mahua butter or mowrah butter, many edible and medicinal applications and it is also used as a biodiesel (Singh and Singh, 1991). The flower has an antibacterial activity against the *Escherichia coli* and resist against rice pest disease (Das and Chaudhary, 2010). In Ayuevedic system of medicine it is stated that the bark of *Madhuca Indica* is useful in the treatment of cancer at the local application (Mills and Bone, 2004). It is a good remedy for itch, swelling, fractures and snake bite poisoning, internally employed in diabetes mellitus (Nishant *et al.,* 2014), fruits are astringent and largely employed as a lotion in chronic ulcer, in acute and chronic tonsillitis and pharyngitis (Patel and Patel, 2011).

Harvesting and Yield

When trees becomes mature, then yield will starts. Creamy coloure, fleshy corolla (flowers) falls in early hour of morning. Mahua yields both flower and seeds. In early morning flowers will fall which should be collected properly. This will continue for 20-30 days. Fruits are ready to harvest by 3rd week of May to 3rd week of June (Dikshit and Kala, 2014; Sikarwar, 2002). Manually they should be collected and put in storage for seed extraction. Yield of dry flowers will be 120-150 kg/tree and kernel will be 70-80 kg/tree. This productivity will continue for 50 years. Flowers before storage are dried. As soon as they are collected they should be dried on polythene sheet. Do not store them in bulk in moist condition. Flowers should be stored in thin layers. Dried flowers after packing in gunny bags are stored in dry place up to 1 year under ordinary condition (Singh, 2007).

According to Singh *et al.* (2006) studied on maturity standard in different genotype of Mahua under Gujarat conditions and observed that fruit growth was faster initially and showed down while reaching towards maturity. TSS, TS and reducing sugar increased as fruits reached towards maturity. Titrable acidity increased initial period of fruit development, then declined. Fruits of MH-1 and MH-5 are ready for harvest by 2nd week of May and 1st week of June respectively.

Value Added Products

Alsi Lata

Sun dried saculent flowers of Mahua (*Mahuca lantifolia*) are generally used for making country wine. In the present study a different preparation called Alsi lata was prepared using seeds of lentil.

Sl.No.	*Ingredients*	*Quantity*	*Sl.No.*	*Ingredients*	*Quantity*
1	Linseed	1 kg	3	Dry fruits	50 g
2	Dry Mahua flowers	500 g			

Cleaned linseed is roasted and grounded. Crushed Mahua flowers are mixed with grounded linseed and medium size balls/laddus/pancakes are prepared. Daily use of Alsi Lata is very energetic to health. It contains 13.99 g protein, 26.27 g of fat, 1.85 g of minerals, 3.23 g of fibre, 26.27 g of fat, 399 K Cal energy, 134 mg of calcium, 257 mg of phosphorus and 2.02 mg of Iron per 100 gram of food (Chauhan *et al.,* 2009).

Seed Oil Extraction

Seeds are good source of oil (Singh *et al.,* 2005). The oil obtained from kernel, which is said to be useful for heart patients is used for edible purpose and permitted for preparation of vegetable oil. Amount of oil obtained from seeds of the fruit is higher than many oil seed crops and oil-bearing trees. In Mahua oil, linoleic and unsaturated fatty acids are found, which are useful for heart patients, because it reduces the cholesterol content in blood serum. Mahua oil is used in manufacture of soap, lubricating grease, fatty alcohols and candles (Dhakar *et al.,* 2015). Mahua seeds contain up to 50 per cent oil. In the industrial extraction process, the seeds are first broken and flaked, the resulting flakes are steam-cooked and the cooked flakes are crushed and solvent-extracted with hexane at 63 ^{0}C. The resulting Mahua oil meal contains less than 1 per cent oil. In smallholder farms, the seeds are only crushed, resulting in an energy-rich Mahua seed cake containing up to 17 per cent oil (Ratnabhargavi, 2013; Singhal *et al.,* 1986).

Ethanol

Mandal and Kathale (2012) studied on ethanol production from Mahua flower using *Sacchromyces cerevisiae*-3044 and maximum production of ethanol is obtained at different optimized parameters such as slurry composition as 1:5, pH at 5 to 5.5, inoculum level at 1.5 g/100mL, inoculum age at 48 hours, temperature 30–32 ^{0}C, nitrogen source 0.05-0.06 per cent, sodium potassium tartarate 1.2 g/L and fermentation period is 2-4 days is 338 ml for 1:5 slurry composition. Maximum production of ethanol is obtained by use of sodium potassium tartarate and urea.

Mahua (*Madhuca latifolia* L.) flower is a suitable alternative cheaper carbohydrate source for production of bio-ethanol (Behera *et al.,* 2010). Mahua seeds contain about 40 per cent pale yellow semi-solid fat. The seed oil is commonly known as "Mahua Butter". The oil content of the seed varied from 33 to 43 per cent weight of the kernel. For the tribals of India, Mahua oil is by far the most important tree seed oil. Fresh Mahua oil from properly stored seeds is yellow in colour with a not unpleasant taste. The oil is used as cooking oil by most of the tribes in Odisha, Chhattisgarh, and Maharashtra *etc.* (Mishra and Padhan, 2013).

As far as Mahua's fuel property are concerned it is comparable with diesel fuel and is available in good quantities that too underutilized. Its calorific value is 96.30 per cent on volume basis of diesel. It does not have any significant change if we mix Mahua oil by 20 per cent in diesel in terms of power output, brake specific fuel consumption and brake thermal efficiency (Kumar *et al.,* 2015). By using Mahua oil blends, performance of engine improved which enhance compression ratio from 16:1 to 20:1 (Rathore and Ramchandrani, 2016). Use of biodiesel would create high job opportunities in developing countries in view of sector of agriculture where currently level of employment is very low (Aregbe, 2010). According to Haiter *et al.* (2012) studied on Mahua oil as substitute for fuel of diesel engine and found that the lower blends (25 per cent, 50 per cent, 75 per cent and 100 per cent) of biodiesel increases the brake thermal efficiency and reduces the specific fuel consumption. The exhaust gas emissions are reduced with increase in biodiesel concentration and it proves that the use of biodiesel (produced from Mahua oil) in compression ignition engine is a viable alternative to diesel.

Mahua Liquor

The flowers of Mahua tree are fermented to produce an alcoholic drink called Mahua, country liquor (Kala, 2011). Dried Mahua flowers are an attractive source of fermented products due to high sugar content. Preparation of Mahua wine from fresh flowers has also been reported (Yadav *et al.,* 2009). Various products like alcohol, brandy, acetone, ethanol, lactic acid and other fermented product has been prepared from Mahua dry flowers (Fowler *et al.,* 1920; Tiwari *et al.,* 1920). The cost of Mahua dried flowers is quit less in comparable to other raw materials source. Tribal men and women consider the tree and Mahua drink as a part of their cultural heritage. They consumed this liquor in all the social gatherings and ceremonies. It is cheaper than IMFL (Indian Made Foreign Liquors) and other country liquors so for a common public, it is quite available, and they consume a huge amount. The freshly prepared liquor has a strong, smoky foetid odour, which disappear on ageing. The most interested thing about the Mahua tree is that it has two fruits in different seasons; the seed oil is extracted from it and used in the several different purposes (Deep *et al.,* 2016). *Madhuca longifolia* has found many important properties yet its several other potential have to be finding out or fully explore by research workers along with the people of tribal community so they may have more and valuable knowledge (Kamal, 2015). Mahua flowers are rich in sugar (68-72 per cent), in addition to a number of minerals and one of the most important raw materials for alcohol fermentation. The maximum yield of ethanol (9.51 per cent) occurred at 25 ^{0}C with pH 4.5 after 14 days of fermentation of Mahua flower juice. The fermented non-distilled alcoholic beverage contained total sugar (8.83 mg/ml), reducing sugar (0.82 mg/ml), total soluble solids (6.37 ^{0}Brix), titrable acidity (0.65 per cent) and volatile acidity (0.086 per cent). Methanol was not detected at any stage of fermentation. The developed fermented alcoholic beverage had characteristics flavor and aroma of Mahua flowers with about 7-9 per cent alcohol (Singh *et al.,* 2013).

Soni and Deys (2013) studies on value-added fermentation of *Madhuca latifolia* flower and its potential as a nutra beverage and found that the blend of Mahua flower with guava at 25 ^{0}C temperature for fermentation improve the flavor of product and has the potential to be presented in the lucrative market for their high antioxidant activity.

Mahua Seed Cake

According to Joshi *et al.* (1990) alkali treated Mahua (*Bassia latifolia*) seed cake (MSC) was supplemented in the ration, clearly showed that the soaking of MSC in a solution of NaOH at 5 and 10 per cent levels (W/W) for 24 hrs resulted in a distinct improvement in the digestibility of protein, utilization of absorbed-N and intake of TDN. The health of these animals was although robust. Their semen characteristics were normal at maturity and on slaughter different vital organs did not reveal any pathological lesions. Incriminating factors present in MSC could thus be detoxified by alkali treatment and the treated MSC could be effectively utilized in the ration of ruminants. Mahua seed cake is relatively rich in protein (16-29 per cent), the solvent-extracted meals being richer. Press cake contains about 5-17 per cent oil while solvent-extracted meal contains usually less than 2 per cent oil. The crude fibre content is generally low, from 3 to 12 per cent. According to Tiwari and Patle (1997) when processed Mahua seed cake, is incorporated in the concentrate mixture and fed with wheat straw, poor quality roughage, the higher allowances of CP have to be provided to lactating buffaloes than the recommended values.

Dular *et al.* (2000) studied on effect of feeding processed Mahua cake on performance of broiler chicks and found that the feed intake, dry matter digestibility, body weight gain, feed conversion ratio, performance index and protein efficiency ratio were significantly ($P<0.05$) higher in control group as compared to 5, 10 and 15 per cent Mahua cake fed groups and no difference of these parameters were observed between 5 and 10 per cent Mahua cake fed groups. The seeds cake is used as anti-inflammatory, antiulcer and in hypoglycemic activity. Bark of *Madhuca longifolia* is very useful for the treatment of diabetes, rheumatism, bleeding spongy gums, ulcer and tonsillitis (Prajapati *et al.,* 2003).

Value Addition of Mahua Flower by Making Pickle

Sl.No.	Ingredients	Quantity	Sl.No.	Ingredients	Quantity
1	Mahua flower	1 kg	7	Kashmiri mirch	10 g
2	Tamarind paste	400 g	8	Asafetida (Hing)	4 g
3	Chilli powder	20 g	9	Fennel seed powder	3 g
4	Fenugreek powder	10 g	10	Salt	140 g
5	Cumin powder	10 g	11	Acetic acid	5 ml
6	Mustard powder	20 g	12	Mustard oil	250 ml

Source: Kumari *et al.*, 2018.

Technological Flow-Chart for the Preparation of Mahua Flower Pickle

Heat oil in pan and add Asafoetida

↓

Add spices and salt

↓

Remove from flame

↓

Add Kashmiri Mirch powder and Acetic acid

↓

Store in Glass Jar and after 6 day it is ready to serve

↓

Add Mahua flower + Tamarind paste and mixed well

Mahua Flower Ladoo

Mahua ladoo is a product providing variety of nutrients and can be recommended for all age group as it has good biological value. The value added product obtained by standardized processing from Mahua flower will help to combat malnutrition to meet the nutritional requirement of growing children while added variety to food.

Sl.No.	Ingredients	Quantity	Sl.No.	Ingredients	Quantity
1	Mahua flower	1 kg	7	Fenugreek powder	10 g
2	Molasses	100 g	8	Dried ginger powder	20 g
3	Ragi flour	500 g	9	Ghee	100 g
4	Cashew nut	100g	10	Salt	140 ml
5	Sesame seed	40 g	11	Coconut powder	200 g
6	Citric acid	4 g			

Source: Kumari *et al.*, 2018.

Technological Flow-Chart for the Preparation of Mahua Flower Laddu

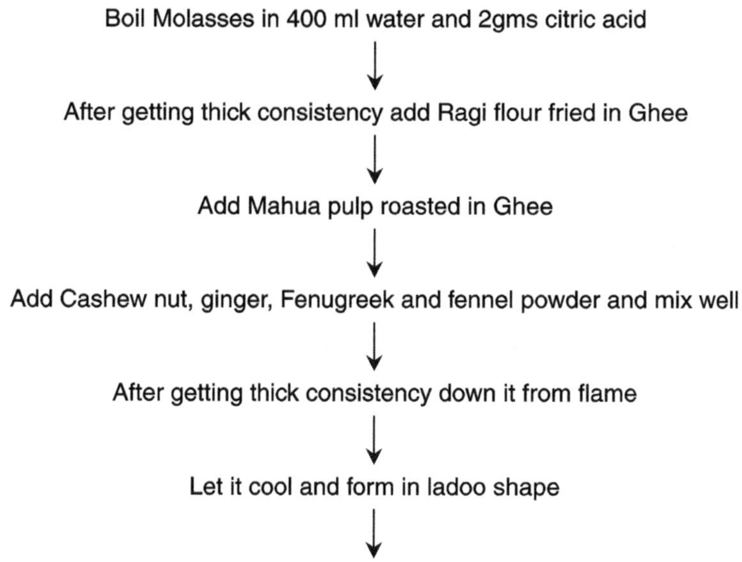

Mahupitthi (Kheer in Mahua Flower Juice)

It is mostely prepared at central and western India such as Andhra Pradesh, Bihar, Gujrat, Madhya Pradesh, Maharastra and Orissa. The flower is rich source of sugar, protein, vitamins and minerals and basically used to make Jaggery.

Sl.No.	*Ingredients*	*Quantity*	*Sl.No.*	*Ingredients*	*Quantity*
1	Mahua flower	200 g	6	Milk	500 ml.
2	Wheat flour	1 cup	7	Almond	8-10 No.
3	Rice (Basmati)	50 g	8	Cashew nut	8-10 No.
4	Jaggery	100 g	9	Dry grapes	8-10 No.
5	Green Cardemom powder	1 tsp.	10	Ghee	1 tsp.

Source: Sahu 2016.

Preparation

Collect Mahua Flower Juice: Soak the dried Mhua flower in one liter of water for 4-5 hrs or overnight, then collect the juice (strained the juice) by using seive or fine cotton cloth.

Preparation of Pithi: Take a wide vessel; mixe wheat flour in small quantity of Mahua flower juice and make dough (if its not so soft then add little more quantity of Mahua flower juice), then makes a very small peas size ball from this dough (Make cylindrical shape and then press slightly).

Technological Flow-Chart for the Preparation of Mahupitthi

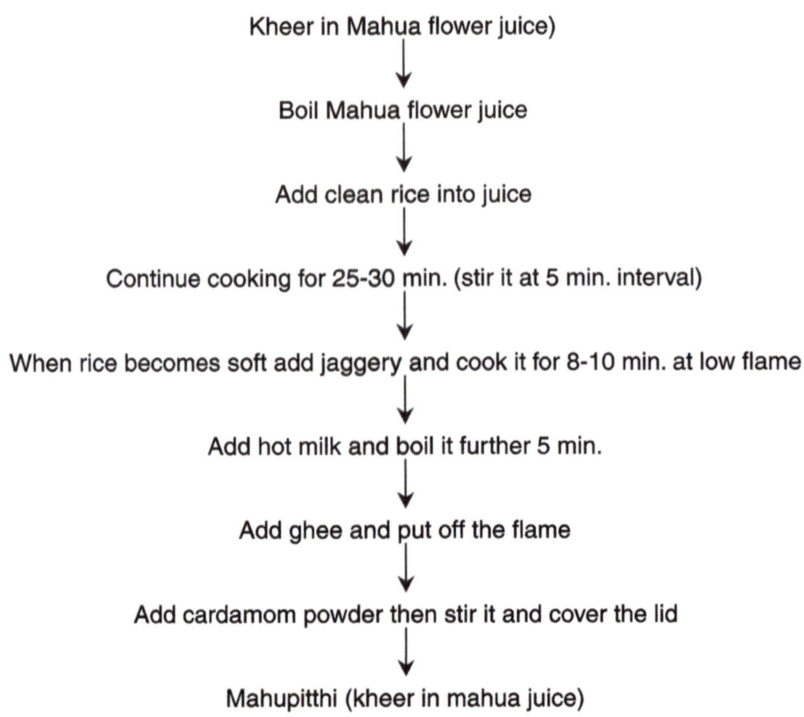

Uses of Mahua Flowers in different Types of Food Products

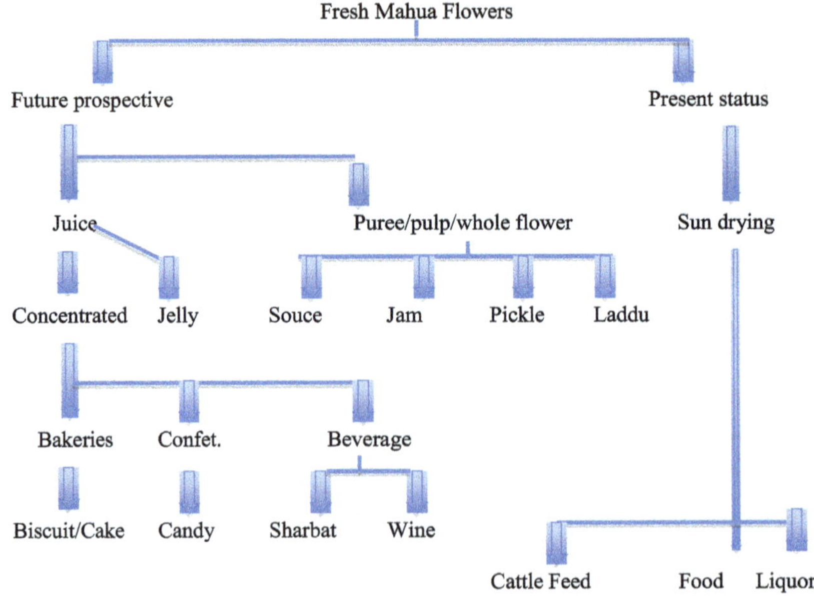

Source: Sinha *et al.*, 2017.

REFERENCES

Anderson, T. M. (2000). Industrial fermentation process. In: *Encyclopedia of Microbiology*, 2nd Ed., Vol 2, (J. Lederberg, Editor-in-chief) pp. 767–781, Academic Press, San Diego, CA.

Anonymous (1995). Wealth of India. In Raw Material, Vol. 6, (L-M) (B.L. Manjunath, ed.) pp. 208–217, CSIR, New Delhi, India.

Aregbe, O. A. (2010). Biodiesel an Alternative fuel for a Cleaner Environment as Compared to Petroleum Diesel. American University of Nigeria Library, TP 359.B46 A231.

Behera, S., Mohanty, R. C. and Ray, R. C. (2010). Comparative study of bio-ethanol production from mahula (*Madhuca latifolia* L.) flowers by *Saccharomyces cerevisiae* and *Zymomonas mobilis.*, 87(7) 2352–2355.

Bina, S. Siddiqui, S. K. and Kardar, M. N. (2010). A New Isoflavone from the *Madhuca latifolia. Natural Product Research.* 24: 76-80.

Chandra, D. (2011). Analgesic effect of aqueous and alcoholic extracts of *Madhuca longifolia* (Koing). *Indian Journal of Pharmacology.* 33: 108-111.

Chouhan, M. P., Singh, S. and Singh, A. K. (2009). Postharvest uses of linseed. *Journal of Human Ecology.* 28(3): 217-219.

Das, B. K. and Choudhary, B. K. (2010). Quantitative Estimation of Changes in Biochemical Constituents of Mahua Flower during Postharvest Storage. *Journal of Food Processing and Preservation*, 34: 831-844.

Deep, P., Singh, A. K., Dubey, S. and Srivastava, S. (2016). Phytochemistry, traditional uses and pharmacological properties of *Madhuca indica* (sapotaceae): A review. *European Journal of Biomedical and Pharmaceutical Sciences.* 3(12): 169-178.

Dhakar, M. K., Sarolia, D. K., Kaushik, R. A., Kumawat, K. L., Singh, S. and Singh, A. K. (2015). Breeding of underutilized fruits crops (*In*): Mahua, pp. 305-323, Ghosh, S. N. (Eds), Jaya Publishing House India.

Dikshit, S. S. and Kala, C. P. (2014). Traditional utilization and harvesting of medicinal plants in Mandla district of Madhya Pradesh. *Applied Ecology and Environmental Science.* 2(2): 48-53.

Dular, R. K., Vaishnava, C. S. V. and Sihag, S. (2000). Effect of feeding processed Mahua cake on performance of broiler chicks. *Indian Journal of Poultry Science.* 35(1): 105-108.

Erik, W. B. and Michael, V. W. (2004). Medicinal Plants of The World, Times Editions, Malaysia, Edition 3: 16-20.

Fowler, G. J., Behram, J. D. E., Bhate, S. R., Hassan, H. K., Mahdihassan, S. and Inuganti, N. N. (1920). Biochemistry of Mahua flowers. *Journal of Indian Institute of Science.* 3: 81-118.

Haiter, L. A., Ravi, R., Arumugham, S. and Thyagrajan, K. (2012). Performance, emission and combustion evaluation of diesel engine using Methyl Esters of Mahua oil. *International Journal of Environmental Sciences*. 3(1): 639-649.

Indian Horticulture Congress (2014). Innovative Interventions by Farmers in Horticulture. An International Meet, Horticulture for Inclusive Growth by 6th Indian Horiculture Congress, November 6-9 at Odissa, pp. 86.

Ingle, P. and Dhabe, A. (2011). Pharmacognostic studies on *Terminalia citrine* (Gaertn) Roxb. Ex Fleming. *Pharmacognosy Journal*. 3(20): 63-65.

Joshi, D., Katiyar, R. C., Sharma, N. C. and Prasad, M. C. (1990). Performance of crossbreed bull calves fed alkali treated Mahua (*Bassia latifolia*) seed cake. *Indian Journal of Animal Nutrition*. 7(4): 245-250.

Kala, C. P. (2011). Indigenous uses and sustainable harvesting of trees by local people in the Panchmrhi biosphere reserve of India. *International Journal of Medicinal and Aromatic Plants*. 1(2): 153-161.

Kamal, A. (2015). Analysis of country made Mahua liquor. *International Journal of Current Research*. 7(12): 23453-23455.

Kirtikar, K. R. and Basu, B. D. (2001). Indian Medicinal Plants. Volume VII Oriental enterprises,: 2058-2061.

Kumar, A., Singh, B. K., Trivedi N. and Agrawal, A. (2015). *Madhuca longifolia*: A review. *International Journal of Pharmaceutical Research*. 1(2): 75-91.

Kumari, K., Sinha, R., Krishna, G., Kumar, S., Singh, S. and Srivastava, A. K. (2018). Sensory and nutritional evaluation of value added products prepared from Mahua flower. *International Journal of Current Microbiology and Applied Sciences*. 7: 1064-1070.

Kureel, R. S., Kishore, R., Dutta, D. and Pandey, A. (2009). Mahua: A potential tree borne oilseed. National Oilseeds and Vegetable Oils Development Board, Ministry of Agriculture, Govt. of India, Gurugaon. pp.12–13.

Mandal, P. and Kathale, N. (2012). Production of ethanol from Mahua flower (*Madhuca latifolia* l.) using *Saccharomyces cerevisiae* - 3044 and study of parameters while fermentation. *Abhinav*. 1(9): 6-10.

Mills, S. and Bone, K. (2004). The essential guide to herbal safety. Elsevier Churchill Livingstone. ISBN: 9780443071713.

Mishra, K. K. and Murty, M. L. K. (2001). A textbook on "Peoples and Environment in India".

Mishra, S. and Padhan, S. (2013). *Madhuca lonigfolia* (Sapotaceae): A Review of its traditional uses and nutritional properties. *International Journal of Humanities and Social Science Invention*. 2(5): 30-36.

Nishant, V., Jha, K. K., Shamim, A., Vipin, K., G., Sudhir, C. and Avnesh, K. (2014). Evaluation of antidiabetic activity of methanolic bark extract of *Madhuca longifolia* in diabetic rats. *The Indian Pharmacist.* 11(12): 31-35.

Owoabi, J., Omogbai, E. K. I. and Obasuyi, O. (2007). Antifungal and antibacterial activities of the ethanolic and aqueous extract of *Kigella africana* (Bignoniaceae) stem bark. *African Journal of Biotechnology.* 6: 882-885.

Patel, M., Pradhan, R. C. and Naik, S. N. (2011). Physical properties of fresh Mahua. *Int. Agrophysics.* 25: 303–306.

Patel, P. K., Prajapati, N. K. and Dubey, B. K. (2012). *Madhuca indica*: A review of its medicinal property. *International Journal of Pharmaceuticals and Research.* 3(5): 1285-1293.

Patel, S. and Patel, V. (2011). Investigation into The Mechanism of Action of *Madhuca longifolia* for its Anti Epilepsy Activity. *Pharmacognosy Communication.* 1: 18-22.

Prajapati, V., Tripathi, A. K., Khanuja, S. P. S. and Kumar, S. (2003). Anti-insect screening of medicinal plants from Kukrail Forest, Lucknow. *Pharmaceutical Biology.* 41(3): 166-70.

Rathore, Y. and Ramchandrani, D. (2016). Mahua Biodiesel: An alternative to conventional diesel fuel. *IOSR Journal of Mechanical and Civil Engineering.* Special Issue. pp 42-51.

Ratnabhargavi, A. R. (2013). Setting up of a Mahua oil extraction unit. Vidarbha Livelihood Forum, Oxfam India.

Rout, G.R. and Das, P. (1993). Micro propagation of (*Madhuca longifolia* Koenig) Macbride var. latifolia Roxb. *Plant Cell Reports.* 12, 513–516.

Sahu, N. R. (2016). Mahupitthi (Kheer in Mahua flower juice): Indian Cooking Mannual; Its all about India Food. http://indiancookingmanual.com/2016/06/mahupitthi-kheer-in-mahua-flower-juice/

Seshagiri, M. and Gaikwad, R. D. (2007). Anti inflammatory, anti ulcer and hypoglycemic activities of ethanolic and crude alkaloid Extracts of *Madhuca indica* Gmein Seed Cake. *Oriental Pharmacy and Experimental Medicine.* 7:141-149

Sikarwar, R. L. S. (2002). Mahua (*Madhuca longifolia* Koen. Macbride)- A paradise tree for the tribals of Madhya Pradesh. *Indian Journal of Traditional Knowledge.* 1(1): 87-92.

Singh, A. and Singh, I. S. (1991). Chemical evaluation of Mahua (*Madhuca indica*) seed. *Food Chemistry.* 40(2): 221-228.

Singh, A. and Singh, I. S. (1991). Chemical evaluation of Mahua seeds. *Food Chemistry.* 40: 221-228.

Singh, I. S. (1998). Mahua an oil bearing tree. Technical Bulletin, pp 3–11, ND University of Agriculture and Technology, Kumarganj, Faizabad, Uttar Pradesh, India.

Singh, P. and Agrawal, O. N. (1989). Detoxification of Mahua (*Bassia latifolia*) seed cake by alcohol treatment. *Biological Wastege*. 29: 229–231.

Singh, R., Mishra, B. K., Shukla, K. B., Jain, N. K., Sharma, K. C. Kumar, S., Kant, K. and Ranjan, J. K. (2013). Fermentation process for alcoholic beverage production from Mahua (*Madhuca indica* J. F. Mel.) flowers. *African Journal of Biotechnology*. 12(39): 5771-5777.

Singh, S. (2007). Underutilized and underexploited horticultural crops: *In*: Mahua. Peter, K. V. (Eds) New India Publishing Agency, New Delhi, India. pp. 283-294.

Singh, S., Singh, A. K., Apparao, V. V., Bagle, B. G. and Dhandar, D. G. (2005). Genetic divergence in Mahua (*Bassia latifolia*) under semi arid ecosystem of Gujarat. *Indian Journal of Plant Genetic Research*. 18(3): 244–249.

Singh, S., Singh, A. K., Apparao, V. V., Bagle, B. G. and Dhandar, D. G. (2005). Genetic Divergence in Mahua (*Bassia latifolia*) under Semi Arid Ecosystem of Gujarat. *Indian Journal Plant Genetic Resources*. 18(3): 244–249.

Singh, S., Singh., A. K., Bagle, B. G. and Dhandar, D. G. (2006). Devlopmental pattern and maturity idices in Mahua (*Bassia latifolia*) genotypes under semi-arid ecosystem of Gujrat. Paper presented in 8th Indian Agricultural Scientist and Farmers Congress, organized by BHU Varanasi and Bioved Research and Communication Centre, Allahabad 21-22nd February, 2006, pp 126.

Singhal, K. K., Chahil, S. M., Sharma, D. D. (1986). Versatile Mahua (*Madhuca indica*) seed cake - A review. *Agriculture Revolution*. 7 (2): 105-123.

Sinha, J., Singh, V., Singh, J. and Rai, A. K. (2017). Phytochemistry, ethnomedical uses and future prospects of Mahua (*Madhuca longifolia*) as a food: A review. *Journal of Nutrition and Food Sciences*. 7(1): 1-7.

Soni, S. and Dey, G. (2013). Studies on value-added Fermentation of *Madhuca latifolia* flower and its potential as a Nutra beverage. *International Journal of Biotechnology and Bioengineering Research*. 4(3): 215-226.

Sujatha, C. H. and Das, P. K. (1988). Evaluation of plant extracts for biological activity against mosquitoes. *International Pest Control*. 30(5), 122–124.

The wealth of India (2007). Raw Material, Council of Scientific and Industrial Research, New Delhi, Vol. 6.

Tiwari, D. P. and Patle, B. R. (1997). Protein requirements of lactating buffaloes fed ration containing processed Mahua seed cake. *Indian Journal of Animal Nutrition*. 14(2): 98-103.

Tiwari, K. P., Pandey, A. and Minocha, P. K. (1980). Influence of aqueous extract of the flowers of *Madhuca indica* on lactic acid formation from molasses by *Lactobacillus bulgaricus*. *Vijanana Parishad Anusandan Patrika*. 23: 61-63.

Varghese, B., Naithani, R., Dullo, M. E. and Naithani, S. C. (2002). Seed storage behaviour in *Madhuca indica* J. F. Gmel. *Seed Science and Technology. 30, 107–117.*

Verma, A. and Singh, U. B. (1979). Techniques of removing saponins from Mahua (*Bassia longifolia*) seed cake and its suitability as animal feed. *Experimentia.* 35, 520–521.

Wayland, C. (2004). The failure of pharmaceutical and the power of plants, medicinal discourse as a critique of modernity in the Amazan. *Social Science and Medicine.* 58, 2409–2419.

Yadav, P., Garg, N. and Divedi, D. H. (2009). Effect of location of cultivar, fermentation temperature and additives on the physiochemical and sensory qualities of Mahua (*Madhuca indica* J.F. Gmel.) wine preparation. *Natural Product Radience.* 8(4): 406-418.

Yadav, S., Suneja, P., Hussain, Z., Abraham, Z. and Mishra, S. K. (2011). Prospects and potential of *Madhuca longifolia* (Koenig) J.F. Macbride for nutritional and industrial purpose. *Biomass and Bioenergy.* 35(4): 1539–1544.

Multiple Choice Questions

Aonla

1. **Best cultivar for preserve (Murabba) making**
   a. Francis b. Kanchan (NA-4)
   c. Banarasi & NA-6 d. None of these
2. **Ascorbic acid (Vitamin C) content in Aonla is (mg/100g)**
   a. 100 b. <100
   c. 150 d. 600
3. **Fruit of 21st century**
   a. Bael b. Ber
   c. Aonla d. Jackfruit
4. **Average number of seeds per fruit in Aonla**
   a. 6 b. 1
   c. 2 d. None of these
5. **Chavanprash and Trifla are prepared from**
   a. Fig b. Phalsa
   c. Mango d. Aonla
6. **Ideal variety of Aonla for pickles making is**
   a. Chakaiya b. Francis
   c. NA-9 d. None of these

7. Number of grades in Aonla is

a. 1 b. 2

c. 3 d. 6

8. Large sized graded Aonla fruits are used for making

a. Candy and preserve b. Chawanprash

c. Trifla d. Pickle

9. Small sized graded Aonla fruits are used for making

a. Candy b. Chawanprash and Trifla

c. Preserve d. Pickle

10. Edible portion of Aonla

a. Mesocarp and Exocarp b. Placenta

c. Endocarp d. All of these

11. The fruit which is consider as wonder fruit for health

a. Bael b. Ber

c. Jamun d. Aonla

12. Aonla fruits are harvested for processing purpose at--days after fruit set

a. 180 b. 80

c. 120 d. 200

13. Astringency may be removed in Aonla processing by

a. Syruping b. Pricking

c. Rubbing d. Brining

14. Which one are not used in Trifla making

a. *Emblica officinalis* Gaertn b. *Terminalia chebula* Retz.

c. *Terminalia bellerica* Roxb. d. *Eugenia jambolana* Lam

15. Richest source of Vitamin C is

a. Aonla b. Bael

c. Citrus d. Orange

Bael

1. Symbol of Lord Shiva is

a. *Aegle marmelos* b. *Syzygium cumini*

c. *Artocarpus heterophyllus* d. *Emblica officinalis*

2. **Richest source of Vitamin B_2 (Riboflavin) is**
   - **a.** Bengal Quince **b.** Holy fruit
   - **c.** Bael **d.** All of these
3. **Which fruit crops takes longest duration for fruit set to ripening?**
   - **a.** Mango **b.** Bael
   - **c.** Jamun **d.** Karonda
4. **Marmelosin found in**
   - **a.** Citrus **b.** Karonda
   - **c.** Jackfruit **d.** Bael
5. The calorific value of bael is
   - **a.** 88 calories/100gm **b.** 59 calories/100gm
   - **c.** 64 calories/100gm **d.** 36 calories/100gm
6. **Bael fruits may be ripened 2-3 months prior to harvest if ethrel are used @**
   - **a.** 500 ppm **b.** 250 ppm
   - **c.** 1000-1500 ppm **d.** None of these
7. **For making preserve fruits are selected**
   - **a.** Tender green (Mature) **b.** Fully ripen
   - **c.** Immature **d.** All of these
8. **KMS and Sodium benzoate is the preservative**
   - **a.** Class I **b.** Class II
   - **c.** a & b **d.** None of these
9. **Sugar, salt and acids are the preservative**
   - **a.** Class I **b.** Class II
   - **c.** a & b **d.** None of these
10. **Major processed products prepared from Bael fruits is**
    - **a.** Preserve **b.** Slab
    - **c.** Panjiri **d.** Jam
11. **Bael fruits are harvested in North India in the month of**
    - **a.** January-February **b.** July-August
    - **c.** November-December **d.** April-May

Ber

1. **Ber fruits are ready to harvest days after flowering**
   a. 150-175 b. 100-125
   c. <100 d. >200
2. **King of Arid fruits is**
   a. Bael b. Ber
   c. Karonda d. Jamun
3. **Poor man's apple fruit is**
   a. Apple b. Pear
   c. Ber d. Plum
4. **Time of harvesting in Ber**
   a. November-March b. April-May
   c. July-August d. None of these
5. Late and heavy production variety of Ber is
   a. Kaithali b. Gola
   c. Mundia d. Umran
6. Ethephon at colour turning stage induces early maturity @
   a. 750 ppm b. 1000 ppm
   c. 200 ppm d. 100 ppm
7. **By the use of unripe fruits of Ber causes**
   a. Vata b. Cough
   c. Fever d. All of these
8. **Ber fruits are mainly graded in**
   a. 10 grade b. 5 grade
   c. 3 grade d. 7 grade
9. ***Zizyphus spp.* Belongs to family**
   a. Rhamnaceae b. Myrtaceae
   c. Apocynaceae d. None of these
10. **Which one of the fallowing will be most suitable for growing under arid zone of India?**
    a. Jackfruit b. Papaya
    c. Sweet Orange d. Ber (Jujube)

Carambola

1. **Carambola also known as**
   a. Star apple b. Star fruit
   c. a & b d. None of these
2. **Which spp. used for pickling purpose**
   a. *Averrhoa carambola* b. *A. bilimbi*
   c. a & b d. None of these
3. **Bearing time in Carambola is**
   a. July to September b. November to January
   c. April to May d. May to June
4. **Carambola belongs to family**
   a. Caricaceae b. Myrtaceae
   c. Anacardiaceae d. Oxalidaceae

Fig

1. **Which tree denotes peace & prosperity?**
   a. Fig b. Aonla
   c. Jamun d. Star Fruit
2. **Which fig produces three crops in a year?**
   a. Smyrna Fig b. Capri Fig
   c. San Pedro Fig d. Adriatic Fig
3. **Harvesting stage of Fig is**
   a. Disappearance of milky latex b. Opening of ostiole
   c. a & b d. None of these
4. **Fig belongs to family**
   a. Rosaceae b. Myrtaceae
   c. Moraceae d. Rhamnaceae
5. **Anjeer is the common name of**
   a. Fig b. Jamun
   c. Mangosteen d. Phalsa

Jackfruit

1. **National fruit of Bangladesh/Poor man's food**
   a. Jamun b. Java plum
   c. Jackfruit d. Mango

2. **Required time from fruit set to maturity in Jackfruit is**
   a. <100 days b. 120-140 days
   c. 275 days d. None of these

3. **Flattening of spines on the rind and thickening of latex is the symptom of maturity indices in**
   a. Grapes b. Mango
   c. Pineapple d. Jackfruit

4. **Full form of JRF**
   a. Jamun rind fibre b. Jackfruit rind flour
   c. a & b d. None of these

5. **Jackfruit belongs to family is**
   a. Anacardiaceae b. Bromeliaceae
   c. Myrtaceae d. Moraceae

6. **Cauliflorus type of fruit bearing is found in**
   a. Jackfruit b. Custard apple
   c. Bael d. Apple

Jamun

1. **Fruit of the god is**
   a. Jamun b. Jackfruit
   c. Mango d. Ber

2. **Fruit seeds are used to control diabetes**
   a. Mango b. Cashew
   c. Karonda d. Jamun

3. **Jamun ripen fruits can be stored at room temperature for**
   a. 2 days b. 6 days
   c. 10 days d. >10 days

4. **Full form of JES**
   a. Journal of Electrical System b. Judgment of ecosystem
   c. Jamun Enriched Shrikhand d. None of these

5. **Jamun belongs to family**
   a. Myrtaceae b. Apocynaceae
   c. Rosaceae d. Caricaceae

6. **Jamun fruit has purple colour due to the presence of**
   - **a.** Carotene
   - **b.** Anthoxanthin
   - **c.** Anthocyanin
   - **d.** Leucoanthocyanin

Karonda

1. **Richest source of Iron is**
   - **a.** Aonla
   - **b.** Bael
   - **c.** Ber
   - **d.** Dry Karonda
2. **Species suitable for jam preparation**
   - **a.** *Carrisa grandiflora*
   - **b.** *Carrisa ovata*
   - **c.** *Carrisa edulis*
   - **d.** All of abobe
3. **Species suitable for scented flowers**
   - **a.** *Carrisa grandiflora*
   - **b.** *Carrisa ovata*
   - **c.** *Carrisa edulis*
   - **d.** Carrisa carandas
4. **Karonda fruit mature after fruit setting**
   - **a.** 100-110 days
   - **b.** 80-90 days
   - **c.** 50 days
   - **d.** 150 days
5. **Karonda is also known as**
   - **a.** Java plum
   - **b.** Christ thorn
   - **c.** a & b
   - **d.** None of these
6. **Harvesting time of Karonda is**
   - **a.** July-September
   - **b.** April-May
   - **c.** January-February
   - **d.** All of above
7. **Most suitable hedge plant is**
   - **a.** Jamun
   - **b.** Mango
   - **c.** Phalsa
   - **d.** Karonda
8. **Karonda belongs to family**
   - **a.** Rhamanaceae
   - **b.** Apocynaceae
   - **c.** Bromeliaceae
   - **d.** None of these
9. **The most suitable plant for bio-fencing is**
   - **a.** Ker
   - **b.** Karonda
   - **c.** Phalsa
   - **d.** Jamun

Phalsa

1. **Edible portion of Phalsa**
   - **a.** Mesocarp & Epicarp
   - **b.** Endocarp
   - **c.** Placenta
   - **d.** None of these
2. **Common name of *Grewia suninequalis* (Syn. *G. asiatica*)**
   - **a.** Phalsa
   - **b.** Dhamani
   - **c.** a & b
   - **d.** None of these
3. **Phalsa belongs to family**
   - **a.** Rosaceae
   - **b.** Tiliaceae
   - **c.** Apiaceae
   - **d.** Myrtaceae

Mahua

1. **Edible portion of Mahua is**
   - **a.** Mesocarp & Epicarp
   - b. Endocarp
   - **c.** Placenta
   - **d.** None of these
2. **Mahua belongs to family**
   - **a.** Rosaceae
   - **b.** Tiliaceae
   - **c.** Apiaceae
   - **d.** Sapotaceae

Answer Key

Aonla									
1	c	2	d	3	c	4	a	5	d
6	a	7	c	8	a	9	b	10	a
11	d	12	c	13	b	14	d	15	a
Bael									
1	a	2	d	3	b	4	d	5	a
6	c	7	a	8	b	9	a	10	a
11	d								
Ber									
1	a	2	b	3	c	4	a	5	d
6	a	7	b	8	c	9	a	10	d
Carambola									
1	b	2	b	3	a	4	d		
Fig									
1	a	2	b	3	c	4	c	5	a
Jackfruit									
1	c	2	b	3	d	4	b	5	d
6	a								
Jamun									
1	a	2	d	3	a	4	c	5	a
6	c								
Karonda									
1	d	2	b	3	c	4	a	5	c
6	a	7	d	8	b	9	b		
Phalsa									
1	a	2	c	3	b				
Mahua									
1	a	2	d						

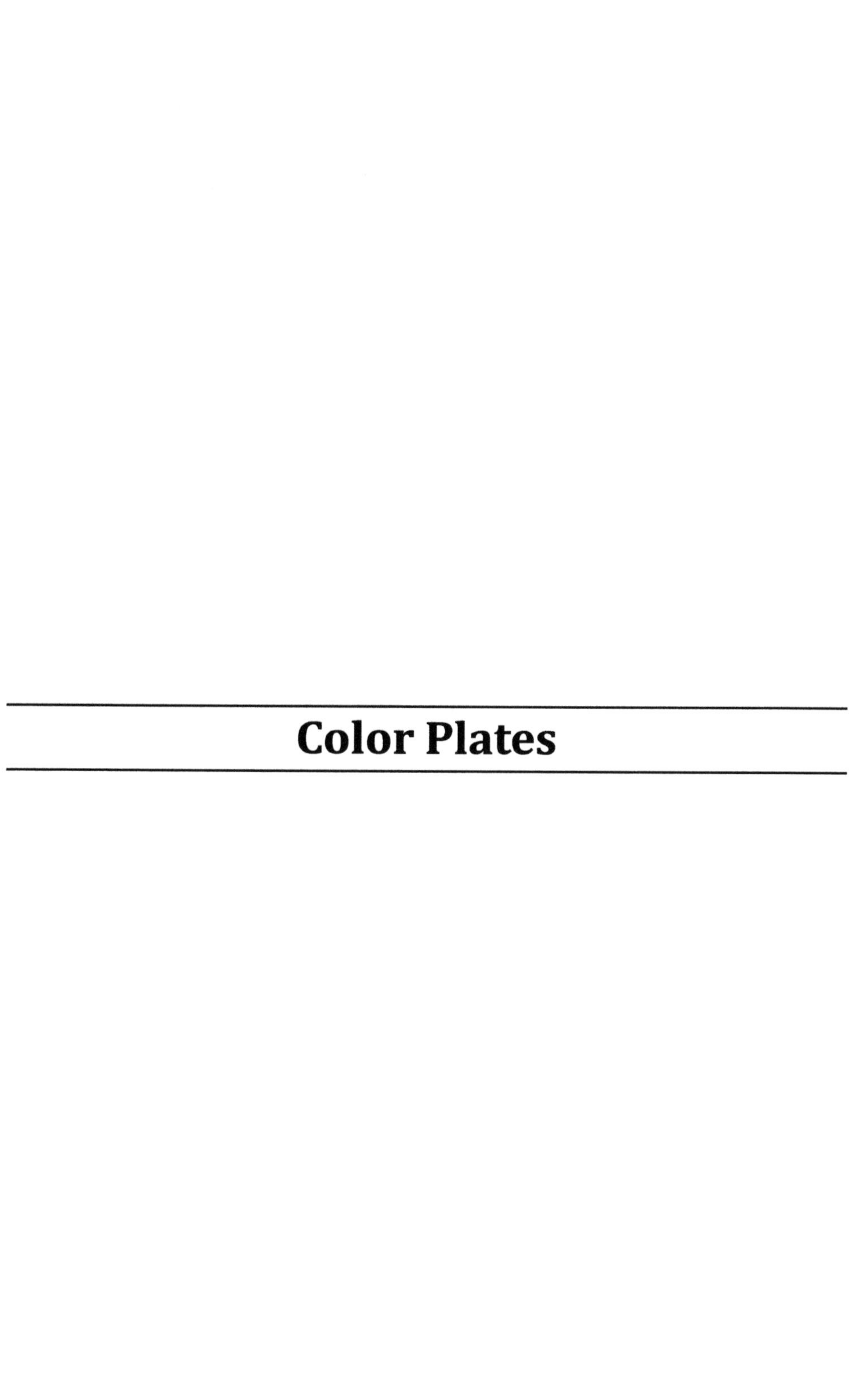

Color Plates

Anola

Aonla jam

Aonla chutney

Aonla preserve

Aonla candy

Aonla churan

Aonla juice

Aonla pickle

Aonla shred

Aonla supari

Aonla chavanprash

Aonla triphala

Aonla biscuit

Aonla mouth freshner

Bael

Bael fruits

Bael preserve

Bael candy

Bael preserve-syrup and bael candy

Bael preserve-syrup

Bael squash

Storage of bael products

Bael sherbat

Bael ready mix

Bael RTS

Bael powder

Bael jam

Bael toffee

Bael pulp

Dehydrated bael

Ber

Packaging of ber

Ber powder

Dried ber

Ber chutney

Ber candy

Carambola

Packaging for overland and sea consignments

Packaging for air freight

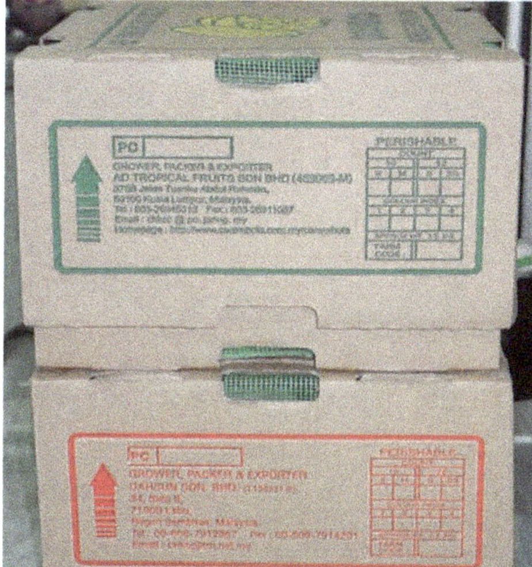

Each box is labeled with information on the origin of the produce
*Note mesh on air holes to prevent post-harvest fruit fly infestation

Star Fruit Jam

Star fruit squash after dilution

Fig

Different stages of fig fruit development, externally and internally

Fresh figs are very sensitive to physical damage, one of the main reasons for their short shelf-life

Fig pulp

Fig jam

Fig nectar

Fig preserve

Fig salad

Fig powder

Fig cookies

Fig toffee

Jackfruit

Jackfruit Brining

Ready to cook tender Jack

Dehydrated jack bulb

Jackfruit bhaji

Jackfruit cutlet

Jackfruit papad

Jackfruit pakodas

Jackfruit pulp

Jackfruit chips

Jackfruit xacuti

Jackfruit halwa

Jackfruit gulab jamun

Jack mini appams

Jackfruit sweet vada

Jackfruit custard

Jackfruit wine

Jackfruit kheer

Jackfruit pudding

Jackfruit fritters

Jackfruit squash

Blended squash of jackfruit + pummelo

Jackfruit RTS

Jackfruit jam

Jackfruit leather

Mango + Jack leather

Jackfruit + Mango sweet

Jackfruit chocolate

Jackfruit seed burfi

Jackfruit seed flour/powder

Jackfruit candy

Jackfruit powder

Jackfruit pickle

Jamun

Blended squash (Jamun + Litchi)

Jamun RTS

Jamun juice

Jamun seed

Karonda

Karonda jam

Karonda jelly

Karonda pickle

Karonda chutney

Karonda candy

Karonda Beverage

Mahua

Mahua processing

Mahua liqour

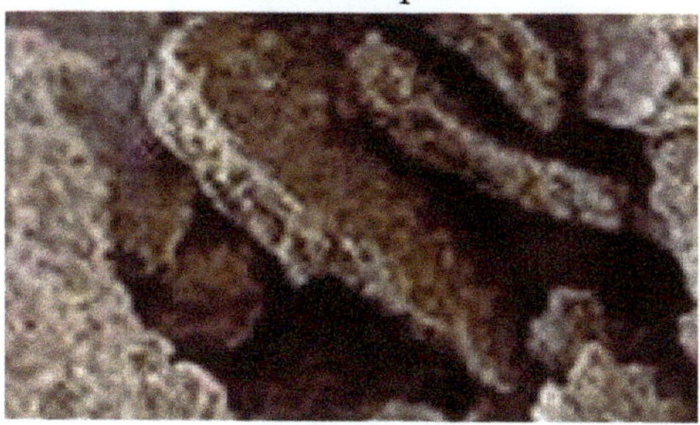

Mahua oil cake

Traditional method of Mahua liqour preparation

Mahupitthi (kheer in mahua flower juice)

Mahua flower ladoo

Phalsa

Phalsa sherbat

Index

www.ingramcontent.com/pod-product-compliance
Ingram Content Group UK Ltd.
Pitfield, Milton Keynes, MK11 3LW, UK
UKHW021010290726
14059UKWH00001BA/62

9 789389 569162